EARTH MATTERS

EARTH MATTERS

How soil underlies civilization

RICHARD D. BARDGETT

OXFORD
UNIVERSITY PRESS

OXFORD
UNIVERSITY PRESS

Great Clarendon Street, Oxford, OX2 6DP,
United Kingdom

Oxford University Press is a department of the University of Oxford.
It furthers the University's objective of excellence in research, scholarship,
and education by publishing worldwide. Oxford is a registered trade mark of
Oxford University Press in the UK and in certain other countries

© Richard D. Bardgett 2016

The moral rights of the author have been asserted

First Edition published in 2016

Impression: 1

Published in the United States of America by Oxford University Press
198 Madison Avenue, New York, NY 10016, United States of America

British Library Cataloguing in Publication Data
Data available

Library of Congress Control Number: 2015944378

ISBN 978-0-19-966856-4

Printed in Great Britain by
Clays Ltd, St Ives plc

Dedicated to Anne Elizabeth Bardgett

PREFACE

A question that I am often asked is: what made you decide to study the soil? Those who ask often do so with an expression of deep curiosity, as if it was an unusual path to have taken. I guess it is, compared to more mainstream fields of science, such as chemistry, physics, or biology. It is also quite a tough question to answer, as there was no single reason why I decided to study the soil. Rather, it was a combination of things, some obvious, some less so. As a child, I was brought up in Cumbria in the north of England, and most weekends we would visit my grandparents at Routenbeck, a small hamlet on the northern edge of the English Lake District. Last summer I went back to Routenbeck, having not been there for several years, and climbed Sale Fell, a small mountain behind the hamlet where we used to play as children. One thing that really struck me was the distinctive smell of the earth, with a deep musky odour that is common to these fells. It took me back to being a child. Having formed in Skiddaw Slate, a friable, old rock that disintegrates into small splinters of slate, the ground had a distinct sound when walked over, which again took me back to my youth. I certainly didn't think about the scientific study of soil as a child, but the smell and sound of the soil left a mark on me from an early age.

My first experience of the scientific study of soil was at school. Geography was my favourite subject. I was fascinated by the land, how it was formed, and how it affected the way humans led their lives. I distinctly remember when I was first introduced to the study of soil. My geography teacher, Eric Rigg, taught us the basics about how soils are formed and how they vary from place to place. For some reason, I was gripped. It was as if I had been introduced to an entirely new world, which most knew little about. A couple of years later, while wondering what to study at university, I stumbled across a Bachelor of Sciences degree in Soil and Land Resource Science at the University of Newcastle upon Tyne, in the north-east of England. Without hesitation, I knew this was the course for me; so I applied, and was accepted onto the course, which I started in 1984. Only a handful of us did the degree; I think seven in total. But it set me off on a career of studying the soil.

It always surprises me why more people are not interested in soil. Many simply see it as dirt or mud, brushing it off their shoes. They consider it to be an odd thing to study or to be concerned about. I am of course biased, but the simple fact is that soil affects our lives in so many ways; in fact, it is of vital importance for human life, and, throughout history, soil has played a major, and often central, role in the lives of humans. Entire societies have risen, and collapsed, through the management and mismanagement of soil; farmers and gardeners worldwide nurture their soil to provide their plants with water, nutrients, and protection from pests and diseases; for centuries, winemakers have prized the soil for the terroir, which creates the special character of their wine; major battles have been aborted or stalled by the condition of soil; an increasing number of artists celebrate the vision of soil; murder trials have been solved with evidence from the soil; and, for most of us, our ultimate fate is the soil. Above all this soil, and the

multitude of organisms that live in it, is the engine for the biogeo-
chemical cycles on which the functioning of the Earth depends.

The reason for the general lack of awareness of soil is hard to
pinpoint, but I suspect it is largely down to the rise of industrial
agriculture and recent shifts of people from rural to urban settings
across the world. As a result, for most members of society, every-
day contact with soil has diminished. But interest in soil is now
starting to grow. Scientists and policymakers worldwide are
becoming increasingly aware of the need to manage soils sustainably in order to feed a burgeoning world population. They are also
looking to the soil for new ways of mitigating climate change,
reducing greenhouse gas emissions from land, and boosting carbon storage in soil. More and more people living in towns and
cities are also turning to the soil to grow their own food in their
gardens and allotments, as they did during the Second World War,
and city planners are starting to consider the benefits of soil in the
design of housing, industrial land, and parks. And to cap it all, the
68th United Nations General Assembly declared 2015 the International Year of Soils, with the goal of increasing awareness and
understanding of the profound importance of soil for human life.
Interest in soil is literally booming.

This book is for those who want to know more about soil and
the many, often unexpected, ways that humans depend on it and
transform it. In doing so, the book doesn't just consider the
present; rather, it delves into the distant past to consider how the
rich tapestry of soils that cover the Earth's surface has been formed
over tens of thousands—or even millions—of years, and how
human relations with soil have shaped, and in some cases destroyed, civilizations of the past. The book also considers the
incredible diversity of life within the soil, and the vast range of

roles that this life plays in maintaining the health, or fertility, of soil. Throughout history, humans have transformed the soil, to the extent that every inch of ground on Earth has probably been affected by our actions in some way. The book considers some of the more obvious ways humans have modified soils, especially through centuries of farming, which has completely transformed soils, in some cases leaving them crippled and in a state of decay. But it also explores some of the less obvious impacts, such as the indelible legacy of war, the hidden, but sizeable, signature of climate change, and the consumption of soil by expanding cities. Finally, the book looks to the future, exploring ways to halt soil degradation and to manage soils sustainably to meet major future challenges, such as food security and mitigation of climate change.

The book is by no means an exhaustive exploration of soils and their relationships with humans. I could have written entire books on each of the main topics covered, as others have, or even on different topics covered within individual chapters. But rather, I wanted to give an overview, or snapshot, of the many ways that they have affected, and continue to affect, our lives. All of our lives are in some way connected to the soil, whether we realize this or not, and my goal here is to simply awaken this awareness and inform readers of the vital importance of soil for human life. Wherever you are in the world, you don't have to look far to see how soil is intertwined with our lives. The food we eat and the wine we drink has a deep association with soil, most likely in a far-flung part of the world. And right now, as I travel on a train into London at high speed, I can see a patchwork of fields growing crops on fertile soil, allotments packed full of vegetables, fuelled by healthy soil, and patches of woodland whose growth depends on the underlying soil. As I enter the city, I can see parklands, with areas of grass that take nourishment from the soil, and sports pitches, where perfect playing conditions go hand in hand with management of the soil. All these soils store vast amounts of carbon, help to mitigate climate change, and are home to a bewildering diversity of soil life, whose activities drive the biogeo-chemical processes on which the functioning of the Earth depends.

My own awareness of soil began as a child, and my fascination with soil has developed through years of working with soil in many parts of the world. I realize that my interest in soil is unusual and that few make a career out of their passion for the soil. But I hope that this book passes on some of the interest that has guided me through my career, and awakens an awareness that the earth beneath our feet matters far more than most imagine.

ACKNOWLEDGEMENTS

There are many people I want to thank for their help in writing this book. I am especially grateful to those who carefully read through the different chapters and provided critical comments that guided me in my writing. For this, I thank my much-valued colleagues David Wardle, Ben Turner, Francis Livens, Heikki Setälä, David Powlson, Ian Hartley, Michael Bahn, Diana Wall, Philippe Lamenceau, and Davey Jones. I am also very grateful to Derek Colquhoun, who proofread the entire book, and an anonymous reader who provided very helpful and supportive comments. Latha Menon and Jenny Nugee of Oxford University Press have been a tremendous help; they guided me through the various tasks of putting a book together and helped me shape my writing for a general audience.

I am also extremely grateful to the many colleagues who guided me in my research, hosted me on field trips, supplied me with papers and images, checked facts, and provided valuable information for different parts of the book; there are too many to mention here, but I want them to know that I am extremely grateful for their help. My daughter, Alice, was an enormous help. Despite having little interest in soil, she spent hours sourcing historic information and literature on many issues related to it, and without her help I would have struggled to put this book together. In writing this book, I have drawn on numerous experiences of working with soils across the world, and I am deeply grateful to all those who, over the years, have passed on to me their knowledge and passion for soil. Without them, this book would not have been possible. There are always people who have a big influence on your work through their support, commitment, or passion for their work. I have been fortunate to work with several people who fall into this category, notably John Whittaker and John Rodwell, and the late Keith Syers, Juliet Frankland, and Gregor Yeates, who all greatly influenced my work and fuelled my conviction to study the soil.

Writing books is a challenge, especially with all the additional demands of an academic job. I have had to juggle my writing with teaching, doing research, applying for grants, giving talks all over the world, attending numerous committees and conferences, and working on advisory boards. My strong desire to write this book and awaken awareness of the importance of soil for humans has kept me on track. But I have also had much support and encouragement, especially from my wife, Jill. She has not only put up with the many hours that I have spent writing and talking about the book but she also read through and scrutinized all of the chapters, giving me frank guidance on whether my writing made sense. For this, I am deeply grateful.

CONTENTS

LIST OF FIGURES

Soil and the Distant Past

Each soil has had its own history. Like a river, a mountain, a forest, or
any natural thing, its present condition is due to the influences of
many things and events of the past.   *Charles Kellogg*

Rainbow Beach is a small town on the coastal dunes of eastern
Australia, near Brisbane. I had travelled there to meet with some
colleagues to sample soils from the vast coastal sand dunes that
surround the area. It might seem an unusual place to visit to
collect soil, but a unique sequence of soils has formed in the
sand dunes, which differ greatly in age. As you move inland
from the sea, the soils get progressively older and deeper, and
more weathered and nutrient-poor. The youngest soils are shallow, having only just started to form in recent sand dunes, whereas
the oldest soils are around half a million years old and can reach 25
metres deep. These are among the oldest, deepest, and most
weathered soils that I have sampled, and what I recall most vividly
is how stunted and sparse the vegetation was that grew there, reflecting their struggle to grow in such ancient, weathered soil. The soils
of Rainbow Beach are by no means the oldest on Earth. Hidden
beneath ice sheets in Greenland, scientists recently discovered a

soil that was 2.7 million years old, a remnant of the fertile tundra that covered the area before the ice sheets came.[1] And scientists working in South Africa recently discovered a soil, now compacted in rock, that is 3 billion years old.[2]

One of the most fascinating things about soil is that it is incredibly diverse; soils vary enormously across continents, countries, and from valley to valley and field to field. Even within a small patch of land, such as a field, forest, or vegetable garden, the underlying soil can vary considerably. Over distances of metres, it might differ in its texture and depth, and in its pH, being acid in one patch of a field and neutral in another. Soils also vary greatly in the diversity of living organisms that live within them. I will go into more detail about the diversity of soil life later in this book; but for now suffice to say that it is vast. Soils also change with time. As I discovered in the sand dunes surrounding Rainbow Beach, ancient soils half a million years old are just a stone's throw away from youthful soils that have only just started to form.

So what causes this variation and why do soils differ from one place to another? This question has puzzled many soil scientists, but the first to tackle it in a scientific way was Vasily Dokuchaev, a Russian scientist widely considered to be the father of soil science. Dokuchaev was born in 1846 and spent many years of his life studying in great detail the soils of his native land, especially the chernozem, a highly fertile, humus-rich 'black earth' that covers a broad belt of land across Russia. Based on his observations, he came up with the idea that the way soil varies from place to place is down to five soil-forming factors: the underlying geology, or parent material; the prevailing climate; the topography; the nature of the vegetation; and the amount of time a soil has had to develop. Fuelled by his observations, Dokuchaev went on to create the

2

world's first scientific classification of soils; he was the literary first to put soils on the map. Inspired by Dokuchaev's idea, others went further, most notably Hans Jenny, a Swiss scientist based in Berkeley, California, who brought mathematics to the problem. In his 1941 publication *Factors of Soil Formation*[3] Jenny proposed a simple mathematical formula to predict how soils form, based on the five soil-forming factors. These soil-forming factors have since guided many soil scientists in their explorations of the soil and still do to this day.

In order to describe a soil, a soil scientist will dig a pit to reveal the different layers, or horizons, of a soil (Figure 1). This is the unit that defines a type of soil, and is usually around 1 metre in width, but can be more. Its upper limit is clear-cut, being the boundary between the soil and atmosphere. But its lower limit can be less obvious, with soil often merging gradually into the underlying bedrock or parent material. Most soils have three or four horizons, although they can have more. The first horizon is a layer of partially decomposed plant remains that lie immediately on the soil surface; this is the organic or O horizon, which varies in depth depending on the type of vegetation that the soil supports, the climate, and the topography of the land; I will say more about this later. Scrape this surface organic material away, and you will find the topsoil or A horizon. This is made up mostly of mineral material that is intimately mixed with organic matter from the surface organic zone; it is the engine room of the soil and a site of tremendous biological activity, being where most microbes and animals live, and where most root growth and nutrient recycling occurs. It is also the most vulnerable part of the soil, being the part of soil most exposed to the activities of humans and the erosive forces or wind and rain. Deeper down the soil profile you will find the B horizon, a

FIGURE I The soil profile is made up of layers, or horizons, shaped by the soil-forming factors. At the top of the soil is the O horizon, a layer of organic matter made of dead plant material. Beneath are the mineral A and B horizon, and the underlying C horizon made up of unconsolidated rock.

distinct mineral subsoil that often contrasts in colour from the surface soil and that beneath it, reflecting the cocktail of chemical reactions and weathering processes that occur in this zone of soil. Finally, at the bottom of the soil profile is found the C horizon, the zone where the soil and underlying bedrock or parent material come into contact. This is the deepest part of the soil.

Dokuchaev believed that the type of soil and arrangement of its horizons within any place was *always and everywhere a mere function* of a particular combination of soil-forming factors. Take the Russian chernozem, with its dark and deep, humus-rich A horizon. These soils formed under the combined forces of grassy vegetation, which produces an abundance of nutrient-rich organic matter, and a cool continental climate that is conducive to rapid humus formation. Moving northwards to the boreal zone, we commonly find the distinctive podzol, with its deep, acidic surface organic horizon, underlain by a bleached layer, termed the eluvial or E horizon, and an orangey-red subsurface B horizon. These soils formed under the combined forces of coniferous vegetation, which produces nutrient-poor litter that is hard to degrade; free-draining, sandy parent material; and cold, wet weather, which is conducive to leaching and the stripping of iron and other elements from the upper bleached horizon and their re-deposition in the underlying subsurface soil. Southwards to the tropics, deeply weathered soils, which can be upward of 50 metres depth, have formed as a result of millions of years of weathering under the hot, humid climate. The soils found here are often deep red in colour due to a build-up of iron oxides, and have little organic matter on the soil surface because plant remains break down rapidly in the tropical climate. These are just a handful of the many types of soil that cloak the Earth's land surface, each with its own distinctive arrangement of horizons, and each having formed under its own particular combination of soil-forming factors.

The main focus of this book is the many, often unexpected, ways that humans depend on soil, both knowingly and unknowingly. But before I explore this, it is first necessary to explain a little about how the soil-forming factors work to shape the soil beneath our feet. I will only give a brief introduction to the soil-forming factors; just enough to inform the reader of the dominant forces that create the enormous diversity of soils on Earth. Also, while Jenny included humans as a biotic factor, alongside the role of vegetation in shaping soil, I consider humans separately as soil-formers. This is simply to emphasize their overwhelming influence, and dependency, on the soil.

PARENT MATERIAL

A quick glance at a geological map of any region of the world will reveal a rich patchwork of geology. Across continents, countries, and even regions within countries, the map will reveal marked changes in both the type and age of parent material reflecting different geological events. This patchwork of geology is the baseline, or starting point, for the development of soil; it is the material from which a soil is made. It might be solid rock, formed millions of years ago from cooling volcanic lava, from skeletal fragments of marine organisms, or eroded particulate material, such as mud, silt, and sand, deposited at the bottom of lakes and oceans and hardened into sedimentary layers. Or it might be material left behind by retreating glaciers or sediments formed by the accumulation of windblown dust, known as loess. This parent material is gradually turned to soil through a battery of weathering processes, driven by the actions of water, wind, and temperature extremes, which gradually break down and shatter rock into finer particles, and a suite of chemical reactions, which decompose minerals. Biological agents of weathering also play their part: plant roots penetrate rocks and crack them open, whereas lichens and

6

microorganisms produce organic acids that dissolve minerals, thereby speeding up the weathering of rock.

Parent material leaves its signature in the soil in many ways. Variations in its mineralogy determine the availability of essential elements in soil, such as calcium, magnesium, potassium, and iron. The soils that form on calcareous rocks, such as chalk and limestone, are rich in calcium, whereas soils on volcanic rocks differ in their elemental make-up depending on the type of rock: soils formed on felsic rocks, such as granite, contain high amounts of silica, but little calcium or magnesium, whereas soils on mafic rocks, such as basalt, are low in silica and rich in calcium and magnesium. Some parent materials even form soils enriched in toxic elements. For example, soils developed on black shales, formed in shallow, stagnant waters, can be enriched in heavy metals such as copper, cadmium, and lead. Also, soils developed on rocks containing metal ores can be a rich source of heavy metals, and serpentine soils, which are derived from ultramafic rocks of very low silica content, especially serpentinite, contain high concentrations of nickel and chromium, which presents challenges for the growth of plants.

Parent material also varies in its grain size, which is mirrored in the texture of soil, the relative proportions of the main soil ingredients: sand, silt, and clay. This is a crucial property for the fertility of soil because it determines its ability to hold onto, or retain, water and nutrients: clay minerals are finer and have a higher surface area than sand and silt, and therefore soils with a high clay content are better able to absorb water and to bind essential nutrients, such as calcium, magnesium, and ammonium, onto their surfaces. Texture also affects drainage of water, with rainwater moving much more quickly through coarse-textured sandy soils than finer-textured clay soils. Because of these differences, soils formed on coarse-textured parent materials, such as sandstones and acidic volcanic rocks (for example, granite), are rich in sand and drain freely, but struggle to retain nutrients and water; these soils are gritty to the touch and easy to dig, but tend to dry out quickly and are nutrient-poor. In contrast, clayey soils, such as those formed from clay-rich marine or lake sediments, limestone, or glacial till, can be heavy and sticky, and hard to dig, but they retain nutrients and water during periods of drought, making them more fertile. Soils formed from silt-rich parent material, such as the deposits of large rivers or estuaries, or windblown dust, also make for fertile soils; they are soft and soapy to the touch, easy to dig, and drain more freely than clay, but being less heavy than clay are more susceptible to the erosive forces of wind and water.

TOPOGRAPHY

A geological map will reveal how the parent material of soil differs across land, but it doesn't show the topography, or shape of the land. It doesn't show the steep-sided mountains and valleys that cover much of the world's land surface, or the rolling landscapes and gentle slopes of the lowlands. While the underlying geology provides the material from which soils are made, the position of a soil within the landscape shapes the way it forms; it affects the local climate, the vegetation that forms, and the flow of water through and across the soil. This is most obvious in mountain areas, where the position of a soil can strongly affect the amount of rainfall and solar radiation it receives, the speed at which water moves through and over the soil, and how susceptible it is to erosion. But it is also important in lowlands. Here, even small variations in slope, or subtle depressions or ripples in the land surface, can influence local climate and water movement, with consequences for the formation of soil.

The starting point for understanding how topography affects the formation of soils is the slope. Much of the Earth's land surface is sloping, and, in general, soils at, or near, the top of a slope tend to be freely drained with a water table at some depth, whereas those at, or near, the valley bottom tend to be poorly drained with a water table close to the soil surface. These differences in drainage strongly influence the way a soil develops: well-drained soils on hilltops have deep, orange-brown subsurface horizons, indicative of an abundance of oxygen, which causes oxidation of iron. As drainage deteriorates towards the valley bottom, the soil profile becomes increasingly blue-grey in colour, a signature of saturated soil and lack of oxygen. In extremely wet valley bottom soils, or in depressions, deep organic horizons can also develop on the soil surface as boggy conditions prevent the breakdown of plant remains. The aspect of the slope can also affect the way a soil develops, with soils facing the afternoon sun being warmer and drier, which aids drainage and the recycling of organic matter. These effects are especially strong in mountains, where aspect strongly influences the exposure of soils to sunshine, wind, rain, and snow.

Another common feature of soils on slopes is that those on ridges and steeper parts of slopes are shallower than those on lower slopes in valley bottoms. This is because of the erosive forces of water, which washes soil particles downslope, feeding the lower soils. I will talk in Chapter 3 about how human activities, and especially intensive farming, have greatly accelerated soil erosion, often with catastrophic results. But it is also a natural process: over time, significant amounts of soil can be moved downslope by water, especially after storms and when vegetation is sparse; and, in extreme cases, erosion by running water can scour gullies into the land causing major soil loss to adjoining rivers, lakes, and even oceans. Moreover, because silt and clay are more easily moved by water than are coarser grains of sand, the soils of lower slopes and valley bottoms tend to be richer in clay and silt, and therefore more fine-textured than those of higher slopes. In a similar way, water seeping downslope through soil carries nutrients and salts, which serves to enrich soils of lower slopes and valley bottoms. Soil also moves downslope under the force of gravity, creating terraces. These terraces are especially common in mountains where steep slopes make the soil unstable and vulnerable to downward creep. In addition, many soils suffer repeated cycles of freezing and thawing, and as soil freezes it expands, making it even more vulnerable to downward creep when it thaws.

CLIMATE

Climate varies enormously across the world: polar regions are marked by deep freezing and much snow; hot deserts suffer extreme temperatures and a lack of rain; and the lowland tropics are hot and humid, with much sunshine and heavy rain. Even over short distances, from one side of a country to another, one valley to the next, or the base of a mountain to its summit, there are steep gradients in temperature and rainfall. This variation in climate has an enormous impact on soil formation because it determines the temperature and moisture content of soil, which governs the speed of chemical and biological reactions that drive rock weathering and breakdown of plant remains—both key processes involved in the formation of soil. Any good student of soil would know, however, that because soils form over thousands if not millions of years, the character of a particular soil that they observe might be a legacy of a past climate rather than the one it experiences today.

As a general rule, for every 10 °C rise in temperature, the speed of chemical reactions involved—the oxidation, hydrolysis, and dissolution of minerals—doubles. The same can also be said for the activities of soil microorganisms and their roles in rock weathering and organic matter decay, which also doubles for every 10 °C, at least up to 35 °C, or down to 0 °C, when most reactions slow down or cease.[4] But it is not just the absolute temperature that matters; it is also the length of time when temperature is optimal for weathering, which is much greater in the tropics than in temperate regions or the poles. And because of the year-round hot and humid conditions, weathering is about three times faster in the tropics than in temperate regions, and around nine times quicker than in the poles.

The influence of climate on soil formation is not just down to temperature: rainfall also plays a major role. As noted by Hans Jenny in his 1941 book,[5] when all other soil-forming factors remain reasonably constant, the organic matter content of surface soil becomes higher as moisture increases. But in many cases it is quite difficult to tease apart the effects of temperature and moisture on soil formation because they change together, often in parallel; as a result, they act on soil together. As we saw earlier, hot and humid conditions in the tropics fuel the chemical and biological weathering, speeding up soil weathering and organic matter decay. By contrast, high in the mountains, the combined forces of cold temperatures and high precipitation act to reduce the activity of soil organisms, slowing down organic matter decay, which causes a build-up of humus on the soil surface. For similar reasons, vast tracts of deep peat have formed in some parts of the world under the combined forces of high rainfall and low temperatures; these conditions have slowed the decay of plant remains and favoured the build-up of great masses of peat over the years.

Rainfall influences soil formation in other ways. As water runs through soil, it strips nutrients and salts, replacing them with hydrogen ions that acidify the soil; a process called leaching. Rainwater also dissolves minerals, especially carbonates that are found in limestone, and transports them deeper down the soil profile, and sometimes out of the soil entirely. Percolating water also washes fine clay particles from the upper soil to lower horizons, sometimes forming clay-rich subsurface horizons. And in arid climates, where rates of evaporation are extremely high, there is a net upward movement of water, which carries with it salts to the soil surface, causing havoc for crops.

THE BIOTIC FACTOR

A major conundrum in soil science is how living organisms fit into the overall scheme of soil-forming factors. Some argue that organisms, especially plants, are the most important soil-formers, so much so that soil cannot form without plants; but others are less convinced. Their reasoning is not that plants don't influence soils, which they undoubtedly do; it is more that vegetation, and other organisms, are themselves governed by climate, parent material, and topography. In other words, vegetation isn't independent. Hans Jenny summed up the problem.[6] He noted:

> It is a universally known truism that microorganisms, plants, and many higher animals affect and influence the properties of soil...the 'mere acting' is neither a sufficient nor essential part of the character of a soil-forming factor.

In other words, to be a soil-forming factor, it must be shown that a change in vegetation, microbial, or animal life can shape the soil while all other soil-forming factors, such as parent material and topography, remain constant. Under such conditions, vegetation acts as a true former of soil.

The Earth is rich with examples of vegetation acting as a former of soil, both in the distant past and the present. When forests first expanded across the Earth's land surface during the Devonian Period, around 400 million years ago, they triggered the formation of soil (Figure 2). Prior to this time, the land surface was sparsely vegetated and plants had not yet evolved true root systems, and surface soils were immature and mainly composed of unconsolidated rock. The expansion of the Devonian forests transformed the soil environment. Tree roots penetrated deep into underlying rock, stimulating its weathering and the liberation of minerals, such as calcium and magnesium, which fuelled the growth of plants. The expanding forests also supplied the soil with large amounts of detritus, boosting soil life and the recycling of the organic matter. Tree roots also formed intimate associations with soil-borne mycorrhizal fungi; these symbiotic organisms helped plants access essential nutrients from the soil, especially nitrogen and phosphorus, and acted as biological agents of mineral weathering, accelerating the formation of soil. This was the period

FIGURE 2 Life on land became more complex as plant communities developed and diversified during the Devonian Period, which triggered the widespread formation of soils.

of the Earth's history when vegetation first started to shape the formation of soils.

Jenny used a more recent example to illustrate how vegetation acts as a true soil-former: the prairie-forest transition of central North America, where native prairie and deciduous forest stand side by side on soils formed from the same parent material, and of similar climate and topography; in other words, all soil-forming factors, apart from vegetation, are the same. Much of this abrupt boundary in native vegetation has now been converted to farming, so only remnants exist. But it once extended across the centre of North America, from Illinois in the north to Texas in the south. The reasons for its existence have puzzled many, some arguing that the prairie is a relic of a past, drier climate of the early Holocene, and the shift to a more humid climate during the middle to late Holocene favoured the expansion of the forest into the prairie. But others have argued that humans have played a role, and that years of grazing by buffalo and repeated fires have kept the forest in check, leading to an abrupt boundary between the two. Whatever the cause, the soils beneath the two vegetation types are strikingly different: prairie soils are much richer in organic matter and darker in colour than forest soils, and are deep and nutrient-rich, being fed by the extensive root networks of productive grassland plants; whereas forest soils have less organic matter and are more leached, and often have a bleached horizon in the subsurface soil.

There are many more examples of vegetation acting as a soil-former; too many to expand on here. But a few are worthy of special mention, such as the differing effect on soils of deciduous and coniferous forests, which I will also touch on later in this book: soils under coniferous forest and heathland show features of the podzol, with its deep, acidic surface organic horizon, or 'mor' humus, underlain by a bleached layer and a red, iron-rich subsurface horizon; whereas those of deciduous forest have a less distinctive organic horizon as plant litter that falls to the ground is quickly decomposed and mixed into the underlying mineral soil, to form a 'mull' humus layer. Soils can also be transformed when new plant species invade native forests. Take the Hawaiian Islands, where the evergreen tree *Metrosideros polymorpha*, locally known as Ōhiʻa lehua, grows on young volcanic soils, which are lacking in nitrogen. These native forests have been invaded by the nitrogen-fixing shrub *Myrica faya* which was introduced to the Hawaiian Islands from the Azores and Madeira in the 1800s as an ornamental or medical plant. Being a nitrogen-fixer, this invader has completely transformed these young Hawaiian forest soils, quadrupling rates of nitrogen cycling and creating nutrient-rich soils that provide a more favourable environment for invasive plants to grow.[7]

TIME

Soils take time to develop and with the passing of time, they change. This soil change is driven by the natural processes of soil weathering and organic matter recycling over centuries and millennia, with soils becoming progressively deeper and, in general, developing more horizons, or layers, as they age. Soils vary greatly in their age. In some parts of the world, such as Africa and Australia, they can be millions of years old. In others parts, such as northern Europe and large parts of North America, they are much younger, having formed in debris left behind after the last Ice Age, which ended around 10,000 years ago. Some soils are even younger, having recently started to form in newly formed parent materials, such as the debris of a retreating glacier, coastal sand dunes, or lava and ash laid down by a recent volcanic eruption. Infant soils are also common in towns and cities, in human rubbish over landfill sites, in construction rubble, and, as I will mention in Chapter 4, in dredged sand and incinerator fly ash.

So how fast does soil form and how long does it take for a soil to reach maturity? This is a question that has puzzled many soil scientists and many estimates exist. Some argue that it takes thousands of years for a mature soil to develop, with clearly defined soil horizons, whereas others believe that they can be fully shaped within as little as a hundred years. Of course, rates of soil formation vary considerably from place to place depending on the other soil-forming factors, such as the type of parent material and vegetation that grows, the slope of the land, and the climate, being faster in humid than in arid parts of the world. But as a general rule, the average rate of soil formation, over the entire surface of the Earth, is around 10 centimetres for every thousand years, or just 0.1 millimetre a year.[8] This means that a soil of 1 metre in depth, as might be found in a nearby field or forest, might have taken 10,000 years to form.

Soils go through different stages of development as they age, from youth to maturity, and on to old age. Infant soils tend to be shallow, with poorly defined, if any, horizons and little organic matter. But as they age, they become progressively deeper; defined horizons develop and organic matter builds up on the soil surface as plants establish and grow. These first stages of soil development can be surprisingly rapid, with horizons starting to show themselves within tens or hundreds of years. But, as I have already stressed, it takes thousand years for a soil to reach its maturity, expressing a clear set of horizons characteristic of a particular type of soil.

Soil formation slows at maturity, but it doesn't stop. In some parts of the world, where ancient land surfaces exist, soil formation has been going on continuously for hundreds of thousands, or even millions, of years. These soils are typically very deep, reaching upwards of 50 metres in depth, and centuries of weathering and leaching have left them depleted in nutrients, especially phosphorus, which progressively declines as soils reach old age.[9] This decline in phosphorus is caused by centuries of leaching, which strips the soil of nutrients. But as soils age, the remaining phosphorus also becomes surrounded, or locked up, by iron and aluminium oxides, which have been produced by centuries of mineral weathering.[10] Because of this, ancient soils not only have less phosphorus, but that which remains is locked up and rendered unavailable for use by plants. The vegetation of these ancient soils, which can be found in many parts of the world, is therefore stunted,[11] and crops are challenging to grow.

MAN AS A SOIL-FORMER

Humans have completely modified large parts of the Earth's land surface, and in doing so have firmly left their stamp on soil. Throughout this book, I will touch on many ways in which humans have had an overwhelming impact on soil, in some cases completely overriding the forces of other soil-forming factors. The most obvious human influence is agriculture: the widespread conversion of natural forests and grasslands to agriculture, and the twentieth-century intensification of farming, have completely transformed the physical, chemical, and biological nature of soils, often with catastrophic results. But humans have transformed soils in other ways too: the spread of towns and cities has consumed expanses of soil under bricks, concrete, and asphalt; past and current industrial activities have left soils scarred and contaminated with a cocktail of toxic pollutants; and, around the world, war has left a lasting and often irreversible scar on soil, leaving it churned, riddled with battle debris, and polluted with metals, dioxins, and oil. Soil is also being transformed by recent climate change, and the loss and gain of plant and animal life, which have both been accelerated by the actions of humans. These are just a few of the many ways in which humans have left, and continue to leave, their mark on the formation of soil.

Human impacts on soil can be surprisingly long-lasting. I touched on war above, which can leave an indelible mark on soil, but historical and even prehistoric land use can also leave its imprint on soil for centuries or even millennia. Ancient irrigation practices of Hohokam prehistoric populations living in Arizona, in the US Southwest, substantially altered the texture of soils, and the legacy on vegetation can still be detected today, some 700 years after the land was abandoned.[12] In Western Europe, effects of farming during Roman times on soil fertility and forest vegetation can still be detected today, almost 2,000 years on.[13] And in England, medieval cultivation has left its legacy in the form of ridge and furrow, which gives pastures a corrugated, undulating appearance, with distinctive patterns of drainage that impact on the properties of soil. Finally, one of the most remarkable soil legacies of the past is terra preta soil, which is found throughout the Amazon. These dark, organic-rich soils have remained fertile to this day, maintaining very high crops yields, but were formed between 500 and 2,500 years ago by native populations who added large amounts of charcoal, organic wastes, excrement, bones, and even pottery, to land.[14] What is remarkable about these soils is that they have been cultivated for centuries, but they remain fertile to this day.

Soils also act as reservoirs of human artefacts, which archeologists use to reconstruct the past. Moreover, these artefacts are strongly linked to the soil-forming factors, which determine the extent to which they are preserved. A good example of this are bog bodies, such as the one found in 1850 by two brothers, Edwin and John Grainge, while digging for peat on Grewelthorpe Moor in North Yorkshire, England.[15] The well-preserved body was dressed in woollen garments and a pair of leather sandals, which led archaeologists to conclude that the bog man dated back to the Roman occupation of Britain. Bog bodies have been found in other parts of northern Europe, some dating back to the Iron Age; but what unites them all is that they are preserved due to a particular set of soil-forming factors that lead to the build-up of cold, wet, acidic, and oxygen-free peat.

Another twist in the tale of soil-forming factors and humans, is that the fingerprint of soil is increasingly being used by forensic scientists to help catch and convict criminals. By analysing the mineralogy or DNA of soil samples taken from the shoes, clothes, or fingernails of a suspect, forensic scientists can link them to the scene of a crime. One of the first examples of soil information being used in a criminal investigation was in 1908, when the German scientist Georg Popp linked the soil on a murder suspect's shoes to that which surrounded the scene of the crime. And even Sherlock Holmes was wise to forensic power of soil. In the mystery *A Study in Scarlet*, Dr Watson commented that Sherlock

> tells at a glance different soils from each other. After walks, has shown me splashes upon his trousers and told me by their colour and consistence in what part of London he had received them.

But now, detectives and prosecutors are increasingly using soil evidence to link criminals to crime scenes, and a battery of techniques are available to help them detect the fingerprint of particular soils.[16]

Wherever you stand, whether it is in parkland, a field, a tropical forest, or in the wilds of the open tundra, the soil beneath your feet will have been shaped by a myriad soil-forming factors. In most cases, these factors will have been shaping the soil for thousands or even millions of years, creating a landscape of enormous soil diversity. Soils vary in their texture, ability to drain water, and availability of nutrients, and they vary in their colour, depth, and capacity to support plants. And while soil growth is painstakingly slow, they are in reality in constant flux; soil is teeming with life, and nutrients, gases, and water are constantly being cycled within soil, and between the soil and their surrounding environment. It is this rich diversity of soil that impacts human life in many, often unexpected, ways. Not only do humans influence the nature of soils, acting as a soil-former themselves, but as you will discover in this book, soils have throughout history also had an overwhelming influence on human lives.

2

●●●●●

Soil and Biodiversity

Nature is often greatest in her smallest creatures   *M. S. Devere*

One of the most striking things about soil is that it harbours a remarkable diversity of life. A handful of soil from any well-kept garden, forest, or agricultural field, can contain literally billions of individual organisms and thousands of species. In some cases, as much as 10 per cent of the soil's total weight could be alive, although in most cases it will be 1–5 per cent. The bulk of these organisms are microorganisms, which aren't visible to the naked eye: the bacteria, fungi, and algae. But the soil is also home to many animals, including microscopic nematodes and protozoa, and large faunas such as springtails, earthworms, spiders, and even moles. The diversity of all these organisms is vast, with some scientists estimating that soils probably contain as much as one-quarter of the living diversity on Earth.[17]

The importance of soil organisms for soil fertility has long been known. The philosopher Aristotle (384–322 BC) referred to earthworms as 'the intestines of the earth', and Cleopatra (69–30 BC), the last pharaoh of Egypt, declared them to be sacred because of their contribution to Egyptian agriculture. Darwin detailed the importance of earthworms for soil fertility in his last book, published in 1881.[18] He commented:

> It may be doubted whether there are many other animals which have played so important a part in the history of the world as have these lowly organized creatures.

Also, the benefits of leguminous plants for soil fertility and crop growth have been known since Roman times. But it wasn't until the late nineteenth century that it was discovered that nitrogen fixation is down to microscopic bacteria (*Rhizobium*) that live in small modules in roots. Around the same time, it was also discovered that bacteria that live freely in soil, outside plant roots, also fix nitrogen from the atmosphere and boost nitrogen supply to soil.

The simplest way to introduce the diversity of life in soil is as a food web (Figure 3), made up of groups of microorganisms and animals with different feeding habits; some feed on dead organic matter, whereas others feed on plant roots and even on each other.[19] Starting at the base of the food web are the primary consumers. These are mainly bacteria and fungi, which are by far the most abundant members of the soil food web, with billions of bacteria and hundreds of kilometres of fungal hyphae within a single handful of soil. Their diversity is equally astounding, with tens of thousands of bacterial and hundreds of fungal species within the same handful of soil. These bacteria and fungi are the factory workers of the soil food web, degrading and transforming complex dead organic matter into more simple nutrients, such as nitrate, ammonium, and phosphate, which can be used by plants for growth. They also perform other important roles in soil: some

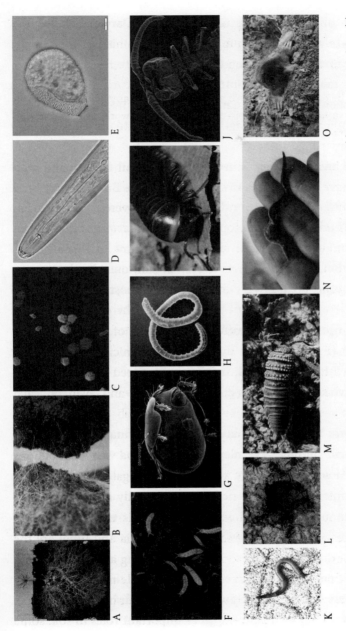

FIGURE 3 A selection of underground organisms that make up the soil food web, including ectomycorrhizal (a) and decomposer fungi (b), ammonia-oxidizing archaea (*Nitrososphaera sp*) (c), nematodes (d), testate amoebae (e), collembolans (f) and mites (g), enchytraeid worms (h), millipedes (i), centipede (j), earthworms (k), ants (l), woodlice (k), flatworms (n), and moles (o).

of them act as pathogens to plants, suppressing plant growth, whereas others produce gums, or glues, that bind soil particles together, thereby improving the structure of soil.

Some microbes form intimate, symbiotic relationships with plants, such as nitrogen-fixing bacteria that live in roots of legumes, and mycorrhizal fungi, which form mutualistic associations with plant roots: these fungi supply essential nutrients, such as nitrogen and phosphorus, to the plant, and in return, the plant gives them carbohydrates to sustain their growth. Moreover, these fungi also form underground networks of hyphae, which serve to connect individual plants to one another, transporting nutrients, water, and even chemical signals to warn of oncoming insect attack, from plant to plant. However, recent research has revealed that plants are actually very choosy about which mycorrhizal fungi they trade with, seeking out those that give them the best deal in terms of nutrient supply, and avoiding those that hoard nutrients rather than paying back the plant for the carbohydrate they provide.[20] In other words, there is a kind of underground biological market, with plants changing their trading partners if they don't get a good deal.

Next are the secondary and higher-level consumers. These are the animals that feed on microbes, plant roots, and even on each other. These animals vary tremendously in size, ranging from microscopic protozoa and nematodes, to medium-sized microarthropods and enchytraeid worms, and finally to the largest animals, which can be seen by the naked eye, such as earthworms, termites, ants, centipedes, spiders, beetles, and even moles. Although not as numerous or species-rich as the microbes, these animals are still astonishingly abundant and diverse. For example, protozoa, which are microscopic, single-celled organisms that live in soil water, can reach remarkable densities of up to 10 million within just a single gram of soil, with perhaps as many as 100 species. Nematodes are also extremely abundant and diverse, with millions of individuals and hundreds of species typically being found in a single square-metre patch of healthy soil. And in the same patch of soil, there might be upwards of 300 earthworms, the best-known subterranean animal of all, although the number of species present will be relatively low.

Very little is known about the diversity of life in soil, at least compared to what is known about life aboveground. So why is this the case? One simple reason is that soil organisms are largely out of sight and mind, and because of this, they attract less attention than the visible diversity of life that lives aboveground. Another reason is that soil organisms are tough to study: not only is it difficult to get them out of the soil, but once out, they are difficult to work with. The majority of microbes in soil can't be cultured in the laboratory, and many soil species are hard to distinguish from one another without a great deal of specialized knowledge and patience. I recall some years ago travelling with a colleague to Wageningen in the Netherlands to visit Henk Siepel, one of the world's few specialists in the study of mites. We went there for his help in identifying some mites taken from a grassland site in Scotland, where we had been exploring how soil organisms are affected by land management. We took with us various test tubes, each with a number of specimens of what we thought were different species of mites. It became clear very quickly that the individuals we thought to be of the same species were in fact different. To us they looked identical, but what we had missed was the microscopic details that distinguish species, such as the positioning of a hair on the body or legs, or the shape of its anal shield. We left Wageningen with the knowledge that there were many more species in the Scottish grassland than we had imagined.

Many scientists are now using molecular tools to get around the problems of identifying soil organisms. Rather than looking down a microscope, it is now possible to extract RNA and DNA from soil and then examine the gene sequences of the myriad species of bacteria and fungi, and even animals, which live in a sample of soil. These methods have revolutionized the study of microscopic organisms in soil, unearthing a bewildering, and often unexpected, diversity of soil life. For example, tens of thousands of bacterial gene sequences are typically found in samples of soil,[21] and even in boreal forests, which are generally thought to be species-poor, the use of molecular methods has unearthed upwards of 1,000 fungal taxa, which is many more than might be expected.[22] Even in places where you wouldn't expect to find much life, such as in hot mineral soils of Mount Erebus, Antarctica, the use of molecular tools has unearthed many novel bacterial sequences, which are relics of ancient microbial phyla adapted to this harsh environment.[23]

A JOURNEY ACROSS THE GLOBE

One of the fascinations of biodiversity is that it is not the same everywhere. Some places are teeming with biodiversity, such as tropical forests and coral reefs, whereas others, like hot deserts and Antarctica, are at first glance devoid of life. Understanding the causes of this extreme variation in the richness of life across Earth has long been a fascination of ecologists. But as is often the case, most have only searched aboveground or in the sea; as a result, much less is known about variation in biological diversity in soil.

In the late 1990s, a couple of soil ecologists, Brian Boag of Scotland and the late Gregor Yeates from New Zealand, got together to try to see what they could learn about how soil diversity changes across the globe.[24] They focused on nematodes, being major enthusiasts for these organisms, but also because they can be found pretty much everywhere on Earth. Having scoured all the literature they could find, the main pattern that they unearthed was that soil nematode diversity was much lower near the poles than in other parts of the world. This wasn't a surprise, given that it is similar to what is found for plants and animals that live aboveground. But what did come as a surprise was that the highest diversity of nematodes was found in temperate regions rather than in tropical forests, which is very different to what is found for plants and animals that live aboveground.

A problem for Boag and Yeates's study was that the available data were very patchy. Most of the information that they could find on soil nematodes was from temperate parts of the world, such as Europe and North America, and they could find very little from the tropics or the poles. This was simply because most scientists working on soil nematodes had done their work in temperate parts of the world rather than in the tropics or the poles. But this created a problem for Boag and Yeates because it meant that their findings were biased towards temperate parts of the world, so their conclusions had to be treated with caution. What was needed to solve this was a thorough study of soils across different parts of the world from the tropics to the poles.

With this in mind, I travelled to Fort Collins in Colorado, USA to meet with Diana Wall, a soil ecologist from Colorado State

University, and Jim Garey, a molecular biologist from the University of South Florida. After a few days of discussions, we came up with the idea of setting out on a journey to sample soils from different parts of the world to properly test how soil animals are distributed across the globe. Over the next two years, we visited and sampled soils from eleven different biomes across the world, including tropical forests in Costa Rica and Peru, arid grassland in Kenya, Mediterranean shrub lands in South Africa, temperate rainforest in New Zealand, prairie grassland in Kansas, shrub steppe in Argentina, boreal forest and Arctic tundra in Sweden and Alaska, and polar desert in Antarctica.[25] With this range of sites, we were able to test how soil animal diversity varies across major biomes of the world, and see whether it peaks in the tropics and then declines towards the poles, as most aboveground organisms do.

Soil samples were taken back to the laboratory and analysed using molecular tools, which allowed us to test how diverse they were in animal life. We found an astonishing level of diversity across the sites, with upwards of 17,000 different gene sequences.[26] But what was surprising was that there was little difference in the diversity of soil animals in samples taken from tundra regions, African or prairie grasslands, or even tropical forests; the only pattern we found was a big drop in the number of sequences towards the poles. The other finding that surprised us was that almost all of the gene sequences detected occurred only at a single location. In other words, the different places we sampled, whether in tropical forest, savannah, or tundra, had a unique set of animal species living belowground.

The studies I have mentioned so far were on soil animals, but how does the diversity of soil microorganisms, the bacteria and fungi, vary across the globe? Only a handful of studies have tackled this question, and it appears that different groups of microbes show very different patterns. In one study, Noah Fierer and Robert Jackson, both from the US, investigated the variation in diversity of soil bacteria across continents.[27] They took soil samples from sites spanning the length of North and South America, and measured the genetic diversity within them by analysing soil DNA. As we found with soil animals, bacterial diversity didn't appear to be related to latitude in any obvious way. Rather, differences in bacterial diversity across sites were most related to soil pH, a measure of the acidity of soil: across North and South America, bacterial diversity was greatest in soils of neutral pH and lowest in acidic soils where fewer microbes can thrive. Soil acidity also explained why the Peruvian Amazon had the lowest diversity of bacteria of all the sites they sampled, which bucks the trend seen for plants, butterflies, birds, and reptiles, which are all much more specious in the tropics than in any other climatic zone. Along with our findings on soil animals, these findings on bacteria suggest that subterranean life doesn't follow the same rules as diversity of plants and animals aboveground.

However, another study revealed a very different pattern for soil fungi, which are one of the most diverse groups of organisms found in soil.[28] Leho Tedersoo and his colleagues carried out a massive study, collecting soils from 365 locations across the world, covering all continents and major biomes except Antarctica. Using molecular tools, they then measured the species richness of soil fungi in their samples and tested how it varied across the globe with climate, soil conditions, and the diversity of plants. Unlike for bacteria, they found that species richness of soil fungi did drop from the tropics to the poles, as it does for most organisms that live aboveground. Surprisingly, however, the rate of decline in fungal richness as they moved away from the equator towards the poles was nowhere near as sharp as it was for plants; in other words, the decline in plant diversity moving from the tropics to the poles was much greater than it was for fungal diversity. This suggests that fungi are better able to tolerate the extreme conditions found in the poles than are plants.

Put together, these studies suggest that patterns of soil biodiversity across the globe are more complicated than might be thought. Some groups of soil organisms, especially soil fungi, seem to follow the same pattern that most aboveground organisms do, being more diverse in the tropics than the poles. But others, such as bacteria and nematodes, appear to depart from this rule. Also, while Tedersoo's study revealed that overall richness of the soil fungi declined from the tropics to the poles, the diversity of some groups of fungi, such as pathogens and mycorrhizal fungi, departed dramatically from this rule. But at the end of the day, what all these studies bring to light is that our knowledge of soil biodiversity around the world is incomplete, and future voyages of discovery are needed to fill the holes.

A WALK IN THE PARK

Most people will be concerned more with the diversity directly beneath their feet, such as in their garden, or a nearby park, field, or forest. This local diversity is shaped by a different set of factors, such as the texture of soil, its wetness and acidity, how stony and deep it is, and which plants grow on the soil. Plants affect soil biodiversity in many ways, the most obvious being through adding large amounts of dead plant litter to soil. Plants shed vast amounts of dead litter each year that falls to the ground, providing an abundant source of food for the decomposer organisms that live in the surface soil. But the nutrient content and toughness of this dead litter varies greatly among plant species, which not only affects the type of organisms that live off it, but also the speed at which it is broken down and recycled in the soil. Deciduous plants, for example, produce nutrient-rich litter that is very palatable to soil organisms, so it is rapidly broken down and recycled in soil, liberating nutrients which are again used by plants. Others, such as coniferous trees and shrubs, produce litter that is nutrient-poor, unpalatable, and often rich in chemicals that are toxic to soil organisms, so it lingers on the soil surface, only slowly breaking down.

This pattern was noticed many years ago by the Danish forester P. E. Müller (1884) while exploring his native Denmark, where he observed that there were two broad types of soils, each with a distinctive type of humus and vegetation types.[29] He called these mull and mor. Mull soils have a surface humus layer that is intimately mixed with underlying mineral soil because of high activity of earthworms and moles, and mor soils have a surface layer made up of undecomposed organic matter which builds up in the absence of earthworms. He also noticed that mull humus forms under deciduous trees that each autumn produce litter that is very palatable to soil organisms, whereas mor humus forms under heathlands that produce litter that is tough and rich in chemicals that make it hard for soil organisms to break it down. This way of classifying soils is still used today: fertile mull soils of deciduous forests and productive grasslands teem with life, and are rich in bacteria, the protozoa and nematodes that feed on them, and earthworms, whereas mor type soils of coniferous forests and heathlands have fewer species, but are rich in fungi and micro-arthropods (mites and springtails), but lack earthworms.

The hidden parts of plants, their roots, also play a role in shaping the diversity of organisms in soil. Roots of different plants vary greatly in the way they grow in soil. Some plants have roots that are very fine and highly branched, allowing them to scavenge a large volume of soil, whereas others have thicker, less branched root systems that take up less space in the soil. Some plants have root systems that penetrate deep into the soil, whereas others only scavenge the surface a few centimetres. These differences in root systems of plants not only have knock-on effects for the physical structure of the soil, changing the habitat for microbes and animals in soil, but roots also pump into soil large amounts of nutrient-rich compounds, called exudates, which also impact on soil life. This is because these exudates, which are rich in sugars, amino acids, and proteins, represent a very high-quality nutrient source for the growth of soil microorganisms. And because plants differ in the amount of exudates they pump into the soil, they have very different effects on the soil communities that use them. The quantity of exudates that plants release into soil can be substantial, with some plants releasing as much as half of the carbon that they take up from their leaves by photosynthesis into the soil. The speed of this transfer of carbon from leaves to roots, and then to soil, is also startling, taking a matter of minutes or hours. In fact, it seems that some plants constantly pump large amounts of these sugary exudates into soil, fuelling the growth of the millions of microorganisms that live there.

Scientists have known for many years that plants differ in the way they affect soils and the organisms that live there. My own interest in this topic started during the early 1990s when

I was working at the Institute of Grassland and Environmental Research, just outside Aberystwyth in Mid Wales. I was working with plant scientists at the time, and we wanted to test the idea that some plant species might boost the growth of microbes around their root systems in order to benefit their own growth, for example by increasing nutrient supply. Our approach was simple: we grew different grassland plants in pots, and when they were mature, we sampled soil from around their roots and analysed it for its microbial communities. As expected, we found marked differences in the structure and activity of the microbial communities around roots of the different plant species, and some plants boosted microbial growth and activity more than others. Also, because no leaf litter fell onto the soil during our experiments, the differences in microbial communities we detected among plant species must have been down to their roots.

We went a stage further and decided that we would cut the plants, to mirror what happens when they are eaten by grazing animals, such as cattle and sheep, or cut for hay.[30] What we found was remarkable: the numbers of soil microbes around roots ramped up in cut plants, which was largely down to fast-growing bacteria, which boost nitrogen cycling and its supply to plants. Follow-on studies by us and others have shown that this is a common response in many plants, with defoliation of most grasses, and even tree seedlings, causing a rapid boost in plant photosynthesis, root exudation, and the growth and activity of microbes in soil around roots. It has also since been shown that this ramping-up of microbial activity following cutting also boosts the cycling of nitrogen in soil and uptake of extra nitrogen by plants, which benefits their growth and nutrition, and accelerates their recovery after cutting. This is an interesting discovery in itself, but more generally what it shows is that plants and soil microbes are intimately linked, and that this relationship benefits the growth of plants.

It is now well known that certain plants in a forest, meadow, or even garden, can shape the diversity of life in soil, and that some plants have bigger effects on soil organisms than others. Legumes, for example, fix large amounts of atmospheric nitrogen and boost nitrogen supply to soil, which has an especially big impact on soil organisms, boosting their growth and activity. Also, plants that can grow rapidly, absorbing large amounts of carbon dioxide by photosynthesis, pump substantial amounts of carbon out of their roots, boosting the growth of microbes and nutrient cycling in soil. And in an opposite way, coniferous and heathland plants produce litter that is rich in compounds such as lignin, that are hard for microbes to break down, which slows rates of decomposition and nutrient cycling in soil. These kinds of effects mean that any change in vegetation, such as might be caused by climate change or the planting of new plant species in a garden or field, will have cascading effects on the fertility of soil.

SOIL BIODIVERSITY AND TIME

The diversity of soil life is not static; it changes constantly with time. The abundance, diversity, and activity of soil communities are constantly under change. This can be very rapid, with big shifts in the growth and activity of soil organisms occurring over a matter of hours, or they can be steady changes that take many years, with soil communities taking different forms over time-scales of tens to thousands of years.

Some of the most spectacular changes in soil life occur over hours or days. The first rainfall after a long drought causes a sudden boost in the growth and activity of bacteria and fungi, causing a flush of nutrient cycling and carbon dioxide release from soil. Likewise, the annual thawing of permafrost soil leads to the resurrection of microbes that were frozen in soil, and a pulse of methane and carbon dioxide release to the atmosphere. Pulses of exudation from plant roots also cause rapid shifts in soil life in the root zone of individual plants, boosting the growth of microbes and animals that feed on them, such as such nematodes and protozoa. As already mentioned, the speed of this process is remarkable: within literally hours, carbon taken up by a plant through leaf photosynthesis can travel to its roots, and then into soil where it fuels microbial growth and causes rapid changes in soil biodiversity.

Soil biodiversity also changes over the seasons. Soil organisms are generally least active during the winter, when soils are cold or even frozen, and plants no longer grow, and most active during the summer months. But this is not always the case. Researchers working in the high mountains of Colorado found that soil microbes are actually most abundant during winter, when ground is covered by snow.[31] Also, in these mountain soils, the make-up of the microbial community shifts completely over the seasons: in winter, fungi that live off dead plant material from the previous year's growth are in abundance, whereas in summer, bacteria that thrive on root exudates are more active. These shifts in the make-up of microbial communities over the seasons also seem to control the supply of nutrients to plants, in that nitrogen that is locked up in microbes over the winter is released in spring after snowmelt, providing a pulse of nitrogen for plants at a time when they most need it. This nitrogen is then returned to the soil as decaying litter when plants die in autumn and locked up by the microbes that hold onto it over the winter. What this suggests is quite

37

remarkable: in the high mountains, and most likely other harsh places on Earth, there is an intimate relationship between plants and soil microbes that is designed to make the best use of a very scarce resource, in this case nitrogen.

The biodiversity of soil also changes over decades and centuries. This is called *succession*, or species change over time. Succession occurs after land has suffered some major disturbance, such as a fire, hurricane, landslide, or volcanic eruption. But one of the best places to study succession is the foreland of a retreating glacier. Glaciers have been retreating worldwide over the past 150–250 years and this often leaves behind an ordered sequence of glacial debris of known age (Figure 4). As you walk away from a glacier, the ground on which you walk becomes progressively older. Because of this glacier, forelands make fantastic natural experiments for exploring how soils and

FIGURE 4 The forelands of retreating glaciers, such as the Rotmoostal (Obergurgl, Tyrol, Austria), offer natural laboratories for exploring how soil life develops over time from sterile beginnings.

plant communities develop over time, from barren moraines at the front of the glacier to well-developed communities at the terminal moraines which can be upwards of 150 years of age. Scientists refer to such sequences as chronosequences—or space-for-time substitutions—and they offer ecologists the opportunity to study ecological processes over time spans much longer than direct observation would allow.

My own interest in studying glacier forelands began when I visited New Zealand some years ago. I was there with colleagues to study how the diversity and make-up of soil communities develops from sterile beginnings. We set off on a trip to the Franz Joseph Glacier on the west coast of the South Island to sample soils from sites of increasing age along its foreland, which spans 120,000 years. I was hooked, and since then my work has taken me to the forelands of several glaciers, including Glacier Bay, south-east Alaska, and the Odenwinkelkees and Rotmoostal glaciers, which are high in the Austrian Alps. Studies on glaciers and other new land surfaces have thrown up a number of general patterns in the way life develops over time in soil. The most obvious is that soil microorganisms become increasingly abundant and active as succession proceeds and soils age. There are also striking changes in the make-up of microbial communities in soil. Initially, they are composed of a handful of bacteria and algae, which fix nitrogen from the atmosphere, thereby bypassing the need to get nitrogen from the soil. But as nutrients and organic matter build up as soils age, they become progressively richer in fungi. Mycorrhizal fungi, which are intimately linked to plant roots, also become more abundant and diverse as succession proceeds, and the members of the mycorrhizal community also change: in early succession, most plants that colonize barren land do not need mycorrhizal fungi to grow, whereas later on, most of the trees and shrubs that grow, after some 50–100 years, depend on mycorrhizal fungi to survive.

There are many reasons why soils become richer in life as succession proceeds, but the most important is simply the build-up of soil organic matter that occurs as soils age. There is virtually no organic matter in the debris left by a retreating glacier or the ash left after a volcanic eruption; they are barren, hostile places for soil organisms to live. But as time progresses, plants and other organisms start to colonize these barren environments, triggering the development of a new living ecosystem. With this comes organic matter, which starts to build up on the surface, and then, progressively, well-developed humus layers and soil horizons begin to form. This organic matter, which comes from the plants that increase in cover as succession proceeds, provides an essential resource on which microorganisms and animals thrive. But there is also a feedback to the plants, as the more abundant and diverse food web of older soils becomes more effective in breaking down this organic matter and recycling nutrients to fuel plant growth. In other words, plant and soil communities work closely together to make succession occur.

Before leaving the topic of succession, I want to mention a study I was involved in some years ago. We were working on newly exposed land at the front of a glacier, which is made up mostly of rock and gravel, and contains virtually no organic matter on which microbes can survive. Because of this it is generally thought that the first life to inhabit barren, newly exposed land consists of algae and lichens, which get their energy from sunlight. When these die, they provide energy for the microbes that decompose this dead plant material, making nutrients available for further plant growth.

A few years ago, I was part of a team of scientists who visited the Odenwinkelkees Glacier, high in the Austrian Alps, to test whether this was really the case. We knew that microorganisms, albeit few, lived in the barren ground in front of the glacier, but we didn't know where they got their energy from. What we discovered was remarkable: there is a surprisingly rich diversity of bacteria and fungi living on newly exposed land that support themselves by feeding on ancient, 7,000-year-old carbon, which predates the glacier.[32] This carbon consisted mostly of 'recalcitrant' plant matter like lignin that is difficult to break down, and while we couldn't pinpoint the source of this ancient organic matter, it is likely a remnant of vegetation from a previous glacier-free period. Whatever the source of the carbon, the finding that a diverse array of microbes appears in barren, glacial terrain before plants do, was a surprise. It's almost as if there's a stage that precedes succession.

DOES SOIL BIODIVERSITY MATTER?

One thing that is becoming abundantly clear is that soil biodiversity is very susceptible to the activities of humans. The conversion of pristine tropical forests to agriculture strips soil of its biological diversity, as does the ploughing of natural grassland. Soil diversity is also threatened by intensive farming; heavy use of fertilizers, overgrazing by livestock, tillage, and pollution of land with agrochemicals can all impair the diversity of soil life. Additionally, climate change threatens the diversity of soil life, especially extreme weather events such as droughts and floods, and also fires, which are becoming more common worldwide. This raises the question of whether the loss of biodiversity in soil actually matters for the functioning of soil.

The question of whether soil biodiversity matters preoccupies many soil ecologists. Any gardener or farmer will know that soil organisms are critical for maintaining a fertile soil, and I have already mentioned some of the crucial roles that soil microorganisms and animals play in maintaining a healthy soil. Earthworms break down dead organic matter and improve the structure of soil; symbiotic organisms, such nitrogen-fixing bacteria, boost the supply of nutrients to plants, and mycorrhizal fungi that live intimately with roots help plants gain nutrients from soil, supplying pretty much all of the phosphorus that they require; the many soil animals, such as nematodes, collembolans, and mites, feed on microbes in soil, and excrete large amounts of nutrients, such as nitrogen and phosphorus, into soil, which then becomes available for use by plants; and soil animals consume and break up plant detritus, mixing it into the soil, and they act as engineers of the soil fabric, creating pores and channels that help water move through soil.

I could go on, but the question I want to tackle here is not whether soil organisms contribute to a healthy soil—they unquestionably do—but more whether changes in the diversity, or the number of species within the soil community, matter. A common view among ecologists is that since the diversity of life in soil is so incredibly high, and because many soil species do similar things, or have similar feeding habits, most of them are probably redundant. In other words, if one species becomes extinct, another one will probably replace it and take over its role. This idea is called the redundancy hypothesis. An excellent example of this comes from the work of Heikki Setälä, a soil ecologist from Finland who specializes in the study of boreal forest soils, which cover much of his country. Some years ago, Heikki, along with his colleague

Mary Ann McLean, set up an experiment where they added different numbers of fungal species to humus taken from a nearby boreal forest.[33] They then measured how quickly the humus decomposed as a measure of soil function. They found that the speed of humus decay increased with increasing fungal diversity, but only up to a point; beyond that, the number of species added made no difference at all. In other words, species diversity of decomposer fungi mattered only when there were relatively few species present, but when the number of fungal species was increased further, it no longer mattered for humus decay. What this suggests is that there is some redundancy among decomposer fungi, and perhaps other groups of organisms that perform similar roles in soil.

Researchers have also looked at how the diversity of mycorrhizal fungi in soil influences plant growth and the diversity of plant communities. Marcel van der Heijden, a soil ecologist from Zurich in Switzerland, did one such study.[34] Along with his colleagues, Marcel set up a number of plots in grassland that differed in the number of species of mycorrhizal fungi that they added. After a period of time, they looked at how the plants responded, and found that the growth and diversity of the plants, and their uptake of nutrients from soil, all progressively increased with greater numbers of mycorrhizal species in soil. In fact, the scale of this response was enormous, with the highest species-richness of mycorrhizal fungi boosting plant growth by an incredible 42 per cent compared to when just one fungus was added to soil alone.

So it appears that soil biodiversity is important for soil fertility and plant growth, but only up to a point, after which the benefits are less clear. However, the picture is not so simple. For example, some soil processes, such as nitrogen fixation and nitrification, appear to be very sensitive to species loss, largely because they are performed by a very small bunch of specialist species. Also, not all soil species are equal when it comes to what they do; some species, the so-called keystone species, matter more for the soil functioning than do others. Because of this, the effect on soil functioning of losing species depends more on which species disappear than on how many are lost. A bunch of studies have shown this, including some of my own. In one study we varied the number of species of collembolans that we added to pots of grassland soil, and measured how the diversity of these animals altered the cycling of soil nitrogen and its uptake by plants.[35] To our surprise, the number of species of collembolans that we added to soil didn't matter. What mattered was which species were there: the presence of some species increased nitrogen uptake by plants, whereas others had no impact at all. In other words, the effect of losing species of soil animals on plant nutrient uptake depended on which species were lost, not on how many were there.

A problem with many studies on soil biodiversity is that they are done under artificial conditions in laboratories. Not only do they use a mere fraction of the diversity found in soils, but the soil conditions bear little resemblance to soils in the field; this makes it difficult to extrapolate findings to the real world. I was recently involved in an experiment with several colleagues from around Europe that attempted to overcome these problems.[36] Our goal was to explore how effects of intensive farming on soil biodiversity influenced the way that soils functioned. To do this, we measured an array of soil life in sites across Europe, in England, Sweden, Greece, and the Czech Republic, and at each site, we sampled soils from different fields that varied in the intensity of modern farming practices. We found that, across all these sites, intensive farming, including tillage and heavy use of fertilizers, had negatively affected the diversity of soil life, and these changes impaired the functioning of soil, making it less able to retain nutrients and cycle them efficiently. These findings clearly demonstrate that soil bio-diversity does matter in the real world, and that the activities of humans that impair soil life also harm the functioning of soil.

The study just mentioned was done on soils transformed by years of agriculture, but researchers working on natural soils come up with similar conclusions. A group of scientists set up a series of experiments across four natural ecosystems, ranging from the subarctic to the tropics.[37] At each site, they manipulated the access of differentsized groups of decomposer organisms to dead plant litter and measured how this affected its decay. This allowed them to test whether the loss of key groups of soil organisms affected the breakdown of litter in a similar way in different parts of world, including the harsh Arctic tundra and forests of topical parts of the world. What they discovered was that across all the sites, the loss of key groups of decomposers slowed litter decay and the release of nutrients into soil, which are key functions on which healthy ecosystems depend. Their findings add weight to the view not only that soil organisms matter for the functioning of natural ecosystems, but also that the loss of key groups of organisms from soils impairs the way they function.

THE ENGINEERS

Before I move on, a group of organisms that are worthy of special mention are the soil engineers, which through their activities transform the architecture of soil, leaving their mark on the hydrology and fertility of soil. The best-known engineers are the earthworms, which burrow through soil, creating networks of pores and channels which act as waterways for the drainage of soil. They can also change the soil to such an extent that they modify patterns of vegetation, benefiting the growth of some plants over others. But there are other engineers that have even more dramatic effects on soil and vegetation. For instance, most ants build large nests that penetrate deep into the soil, which not only transform the hydrology of soil, helping with soil drainage, but also create hotspots of nutrient availability, which support lush patches of vegetation above the nests. Other ants engineer their nests on the soil surface, which also create hotspots of nutrient availability that transform the vegetation aboveground. These effects can be dramatic. For example, the massive occurrence of anthills in mountain grasslands of Slovakia has created a runaway feedback that has accelerated their conversion to forest. Here, mounds of the yellow meadow ant (*Lasius flavus*), one of the most common ants in central Europe, promote the colonization and growth of spruce seedlings, accelerating succession of grassland to spruce forest.[38]

In tropical parts of the world, the termite is the main engineer of soil. These small, highly abundant, and species-rich insects create vast mounds and underground galleries in which they live, culti- vate food in 'fungus gardens', and house the egg-producing queen (Figure 5). Termite mounds are common in Africa, Australia, and South America, where they can reach upwards of 5 metres in height, being formed by the movement of many tonnes of soil. These mounds and underground galleries not only transform the soil in their immediate vicinity, but they also shape the entire landscape. In African savannah, for example, termite mounds create a striking pattern that can be seen from the air, with small

FIGURE 5 Termites move huge amounts of soil to build mounds, which transform architecture and hydrology of soils.

dots, or islands of plant growth, scattered across the landscape. These islands of fertility are more nutrient-rich and better able to hold water than surrounding soil, and they support lusher vegetation which attracts an abundance of insects and other plant-feeding animals, such as zebra and buffalo; in doing so, they act as 'supermarkets of the savanna'.[39] New research has also unearthed another role for termite mounds, buffering the effects of future climate change. Working in Africa, scientists showed that by acting as refuges for plants, termite mounds make savannah ecosystems more robust and better able to withstand the forces of climate change.[40]

THE GOOD, THE BAD, AND THE UGLY

Not all soil organisms are beneficial. The soil is a refuge for many pests and diseases that can cause considerable harm to plants and humans. Many soil fungi attack plants, and, as any farmer will know, there is an exhaustive list of soil-dwelling bacterial and viral diseases, and also invertebrate pests, that frequently devastate crops. There is also an endless list of sources of human diseases, including bacteria, fungi, viruses, and protozoans, that spend some part of their life in soil, with major consequences for human health. The World Health Organization estimated that the soil-borne bacterium *Clostridium tetani* is responsible for almost half a million deaths to newborns and 50,000 mothers each year due to tetanus, and billions of infections and over 1 million deaths occur each year from soil-borne parasitic worms and protozoa. I could go on, but the key point here is that human health is strongly affected by pathogens in soil, both directly through disease, and indirectly through their catastrophic effects on crop yields.

Soil pathogens can also affect the diversity of plant communities through a process called negative feedback: growth of a particular plant species for year on year causes a build-up of pathogens around its roots, which curtails its growth more than those other plants growing around them. This is well known to farmers, who use crop rotations to prevent the build-up of underground pathogens of their crops, which otherwise would devastate crop yields. But plant ecologists are beginning to realize that this also occurs in natural ecosystems where it can cause vegetation change. A classic example comes from the sand dunes of the Netherlands. Here, Wim van der Putten and his colleagues discovered that the build-up of pathogens around the roots of Marram grass (*Ammophila arenaria*), the first plant to colonize the sand dunes, eventually leads to its demise and replacement by another grass, *Festuca rubra*, that isn't susceptible to the pathogens. As a result, the vegetation of sand dunes changes, becoming richer in *Festuca* as time passes by.[41]

Scientists have also discovered that soil pathogens play a key role in encouraging plant diversity. In grassland, for example, a build-up of pathogens prevents some plant species becoming too dominant and overwhelming others, which relaxes competition between plants and allows more species to live alongside each other.[42] In a similar way, underground pathogens help explain the relative abundance of different tree species in tropical forests. Studies reveal that adult trees of these forests harbour underground pests and diseases that harm seedlings of their own species more than they do seedlings of other tree species.[43] Moreover, the harmful effects of underground pathogens are much greater for seedlings of rare species than for seedlings of common ones. What is even more remarkable is that this kind of 'self-inhibition' actually explains how common or rare different tree species are as adults in tropical forests, with those that suffer most when growing in their own soil being most rare.[44]

The story doesn't end there, as soil pathogens are also thought to play a key role in explaining why some forest species are able to invade into forests in new parts of the world. A key reason why the black cherry tree (*Prunus serotina*) has been so effective in invading European forests, for example, is because the soil pathogens that keep its growth in check in its native North America do not exist in Europe.[45] Also, by escaping from its pathogens, it grows more aggressively, suppressing the diversity of the native plant community. Together, what all these examples show is that the answer to how plant communities are arranged often lies underground.

Like a great city, the soil is alive with the vigorous activity of its citizens. Beneath every footstep, whether in a park, forest, grassland, or high mountain plateau, the soil is teeming with life. Moreover, these citizens of the soil perform many vital roles that maintain the fertility of soil and the health of terrestrial ecosystems. They break down organic matter, cycle nutrients and carbon, and help to form good soil structure, which collectively sustain the growth of plants and the animals that feed on them. Soil organisms also act as engineers or architects of soil, building physical structures that completely modify the hydrology and vegetation of terrestrial ecosystems. As scientists look more deeply into the soil, they are discovering new roles that soil organisms play, but also recognizing that soil organisms are highly sensitive to the actions of humans: intensive farming, deforestation, soil pollution, tillage, and climate change all reduce the diversity of life in soil to varying degrees. And any human activity that strips the soil of its citizens and disrupts the web of ecological interactions will likely impair the way that ecosystem performs.

3

Soil and the Grower

The nation that destroys its soil destroys itself. *Franklin D. Roosevelt*

Throughout history few things have mattered more to humans than their relationship with soil. This is a bold statement, but since the dawn of civilization, the use of soil to grow crops has been of central importance to mankind. Not only have past civilizations relied on fertile soils to fuel their prosperity and growth, but also neglect of soil, leading to its degradation, has in many cases led to their collapse. Unfortunately, soil neglect isn't just a thing of the past. Today, at a time when the need to produce food for a growing world population couldn't be more acute, vast tracts of once productive land lie degraded and struggling to grow crops or support livestock. The causes of soil degradation are complex, with population growth, poverty, poor delivery of information to farmers, conflict, shortage of land, and climate change all playing a role. Whatever the cause, the consequences are the same: soil degradation causes food shortages, poverty, and hunger.

But I don't just want to discuss the neglect of soil and its dire consequences for mankind. I also want to consider what makes a soil fertile and able to support healthy crops year after year, and how soil can leave its fingerprint on the quality and taste of what we grow. I also want to consider some of the ingenious ways that humans have devised to maintain soil fertility and boost crop yields, and how this knowledge can be harnessed to restore degraded soils. Before I do, however, an important point to make is that the concept of soil fertility is largely agronomic: it relates to the ability of soil to sustain the growth of agricultural crops through the continued provision of nutrients, water, and anchorage. Soils that support some of the most valued natural habitats on Earth, such as pristine tropical forests or expansive arctic tundra, are very infertile from an agricultural point of view; tropical forest soils are highly weathered and nutrient-poor, whereas tundra soils are wet, acidic, and low in plant-available nutrients. But plants have evolved some quite remarkable ways of coping with such conditions, and because of this, these soils have sustained plant growth for thousands of years, and will continue to do so unless humans intervene. To put it plainly, while natural soils might be considered infertile from an agricultural point of view, they are not of ill health.

So what makes a soil fertile? Soils vary greatly in their fertility, with some being naturally more fertile than others. The hallmarks of a fertile soil are many. It will be friable, porous, and well aerated, breaking easily into individual aggregates with a gentle touch of the hand—the signs of good soil structure. A fertile soil will also be rich in colour, showing signs of good aeration and drainage, and be of neutral pH, which is optimal for the growth of most plants and the microbes that transform nutrients into plant-available forms. It will also contain a good amount of nutrient-rich organic matter, or humus, that is well mixed with mineral soil. This organic matter binds soil particles together, helping create good soil structure, and holds onto water, making soil effective at retaining moisture at times of drought. Finally, as I discussed in Chapter 2, it will be teeming with life, with a rich diversity of microbes and animals whose activities promote organic matter breakdown and nutrient recycling, and formation of good soil structure.

A major conundrum in soil science is how to measure soil health. Soil scientists have come up with many ways of doing this, such as measuring plant-available nutrients, the physical structure of soil, or the activity of soil life. But a problem with all of these is that no single measure defines soil health; rather it depends on a rich web of physical, chemical, and biological factors that together operate to give a soil good health. I often joke with students that you don't actually need sophisticated tests to see if a soil is fertile or not. An experienced soil scientist can tell much about a soil from its look, smell, and feel; fertile soil will break away easily from your fingers into well-formed aggregates, it will be rich in colour and smell, indicating well-oxygenated conditions and good drainage, and its organic matter will be intermixed with underlying mineral soil due to high biological activity. This might be a very unscientific view, but simply digging a hole and looking at a soil reveals much about its health. You can see if it has good aggregate structure, whether its drainage is impeded, or whether it is compacted at depth. Of course, simply looking at soil doesn't tell you about the concentration of nutrients or pollutants, or about the activities of microorganisms that are essential for a fertile soil. But it does give some signals about the fertility of the soil.

EARLY IDEAS ON SOIL FERTILITY

Ever since agriculture began, some 12,000 years ago, people have recognized the importance of soil for growing their crops. Ancient civilizations of the Middle East, Europe, South America, and Asia prospered on fertile soil, and as agriculture and populations grew, so did the challenge of maintaining soil fertility. To combat declines in soil health, ancient Greek philosophers advised farmers on the importance of adding manures to their soil and the hazards of soil erosion. The Romans understood that their wealth depended on soil; they identified different soil types on which crops grew best, devised simple ways of testing the quality of soil, and recognized the importance of crop rotations, irrigation, drainage, and manures to maintain the fertility of soil. Ancient Chinese texts contain detailed descriptions on the characteristics of soils, which paved the way for elaborate soil classification systems. And with no comprehension of the process of atmospheric nitrogen fixation, the world's first farmers recognized the important role of legumes, such as beans, lentils, and chickpeas, in boosting the fertility of soil.

As agriculture expanded in the seventeenth and early eighteenth centuries, agriculturalists professed the importance of adding lime and manures to soil, and of using cover crops and crop rotations to nourish the soil. Many also argued that the key to a healthy soil was the right balance of arable and grassland, which allowed for the regular return of cattle manure to fields. But around this time, the first glimmers of scientific soil knowledge began to come to the fore. As early as the mid-seventeenth century, Fellows of the Royal Society in London mused over the thoughts of John Evelyn on soil formation and the potential for chemicals to fertilize the land;[46]

and in 1845, Charles Daubeny published a memoir in *Philosophical Transactions of the Royal Society*, the world's first scientific journal, on the chemical make-up of soil and uptake of minerals by crops.[47] Also around this time, a major scientific breakthrough occurred, which laid the foundation for a transformation of the agricultural world: the German chemist Justus von Liebig discovered that plants required mineral nutrients—including nitrogen, phosphorus, and potassium—for their growth, and that crop yield was proportional to the amount of the most limiting nutrient, whichever nutrient it might be. This is what he called the 'Law of the Minimum', which in simple terms means that if the deficient nutrient is added to soil—be it nitrogen, phosphorus, or potassium—crop yields will increase until another nutrient starts to limit crop growth. These ideas would later transform the agricultural world.

Also active at this time was John Bennet Lawes of Rothamsted Manor, England, who had started to experiment with producing fertilizers, which could be added to soil to boost crop yields. He mixed crushed animal bones, which are rich in phosphates, with sulphuric acid; this generated a compound called 'superphosphate of lime', which he found to be a very effective fertilizer. Previously, some farmers had applied untreated crushed bones to their land, but their effect on crop growth was variable because the phosphorus contained in them is only slightly soluble. Lawes found that treating crushed bones with sulphuric acid greatly increased the solubility of phosphorus and its availability to crops. He put his finding into manufacture at his factory on the River Thames in 1843, filed a patent for the process, and in doing so set in motion the fertilizer industry. Lawes didn't stop there. In partnership with the chemist Joseph Henry Gilbert, he set up a series of innovative experiments at Rothamsted[48] in southern England, to compare how different manures and fertilizers affect crop yields. The most famous of these was the Broadbalk experiment, which was set up in 1843 to see how wheat yields responded to manure and fertilizer treatments (Figure 6). They also established other experiments to investigate ways of improving hay yields on permanent pasture, and yields of spring barley and root crops. These experiments, most of which still run today, not only paved the way for modern scientific agriculture, but also helped establish the principles of managing soils for sustained yields.

FIGURE 6 Harvesting the Broadbalk Experiment in the 1890s. This experiment was set up in 1843 at Rothamsted Manor, England, by Sir John Bennet Lawes and Joseph Henry Gilbert to see how wheat yields responded to different manure and fertilizer treatments. The experiment still continues today.

As the number of scientists becoming interested in soil grew, the issue of soil fertility became a matter of considerable debate. Some, such as Liebig, argued that the soil's ability to continuously support plant growth depended simply on a supply of the limiting nutrients. Others, such as Milton Whitney, the chief of the US Department of Agriculture's Bureau of Soils, disagreed. In a bulletin published by Whitney and his colleague Cameron, it was argued that soil fertility depended more on its physical condition, including its texture and ability to retain moisture, than its chemical composition.[49] Whitney went further in 1909, arguing that soils are an inexhaustible source of plant nutrients, stating:

> The soil is the one indestructible, immutable asset that the Nation processes. It is the one resource that cannot be exhausted; that cannot be used up.[50]

Unsurprisingly, this view created much opposition, especially from Eugene Hilgard, an eminent US soil scientist at the University of California. Hilgard argued that Whitney's experiments were flawed and his ideas unfounded, and also that physical, chemical, and biological characteristics collectively determined soil fertility, not just one or two of them.[51]

Debates rumbled on. But ideas on soil fertility began to take a more biological track following the discovery of the German chemist Hermann Hellriegel in 1888 that legumes, such as alfalfa, clover, peas, and beans, could fix nitrogen from the atmosphere. He also discovered that they did this through microorganisms that converted atmospheric nitrogen to ammonia in root nodules. Around the same time, the Dutch microbiologist Martinus Beijerinck also discovered the bacteria involved, which he placed in the genus *Rhizobium*. Farmers had long held legumes to be important soil improvers, and at the time, they were widely used in crop rotations to fuel the soil with nitrogen; but no one had any idea how they worked. Now, equipped with this new knowledge, soil scientists could begin to understand the soil conditions that maximized nitrogen fixation and how best to harness this knowledge to boost soil fertility and crop yields.

The world's population was growing rapidly during the turn of the late nineteenth and early twentieth century, and to meet the increasing demand for food, the call for new sources of nutrients to fertilize soil and boost yields heightened. The world's phosphate reserves, such as guano and rock phosphate, were already being mined to exhaustion, so people started to look to science for a solution. Sir William Crookes, the president of the British Association, was among them. He told the Association in 1898 that to feed the world's burgeoning population required a new way of fertilizing the soil, tapping into the unlimited supply of nitrogen in the atmosphere to convert it into a form that could be used by crops. He argued: 'It is the chemist who must come to the rescue.'

Just ten years later in 1909, his call was met: the German chemist Fritz Haber discovered how to capture atmospheric nitrogen and produce ammonia. A few years later, another German, Carl Bosch, turned this process into commercial production, now known as the Haber–Bosch process. This was the birth of the artificial nitrogen fertilizer industry, which marked a shift towards a chemical view of fertile soil.

THE GREEN REVOLUTION

The use of synthetic fertilizers boomed during the twentieth century, leading to extraordinary increases in crop yields. This was especially marked from the 1960s to the turn of the century, the period known as the Green Revolution. During this time, the world's average wheat yield ramped up from 1 tonne per hectare to around 3 tonnes per hectare; yields of other staples, such as rice and maize, also boomed.[52] Several agricultural innovations contributed to this increase in crop yields, including the breeding of high-yielding crops; the use of pesticides to control pests, diseases, and weeds; and the introduction of new farm machinery. But the use of synthetic fertilizers probably had the biggest effect, transforming agriculture and also the way farmers viewed the soil. However, it should be noted that these improvements have not been enjoyed everywhere; the Green Revolution bypassed much of Africa, and whilst wheat yields in Europe and North America boomed, they remained meagre in many African countries.

The benefits of intensive agriculture for global food production are unquestionable. But they came at a considerable cost for the soil, the biggest being the loss of organic matter, the staple of a fertile soil. Soil organic matter not only nourishes plants, but also absorbs water and binds soil particles together, improving soil structure and the ability of roots to forage the soil. It also fuels the myriad microorganisms and animals that live underground, the factory workers of a fertile soil. Declines in soil organic matter go hand in hand with intensive farming. Not only does continuous tillage break up and aerate the soil, increasing the speed that microorganisms break down organic matter, but the use of bag fertilizer to nourish crops, rather than organic manures or legumes in crop rotations, results in less organic matter entering the soil, which exacerbates organic matter decline. Synthetic fertilizers don't destroy soil organic matter; rather, they simply replace the use of manures and cover crops, which reduces the amount of organic matter entering soil.

You don't have to look far to see the problems that this can cause. Years of continuous cultivation of the US Corn Belt, the breadbasket of America, has left soils depleted in organic matter, compacted, and susceptible to erosive forces of wind and water. The resulting soil erosion can be considerable, with literally billions of tonnes of topsoil being swept away from US croplands every year.[53] The picture is no better in Europe, where large amounts of soil are regularly washed away from arable land following heavy storms, and in China, centuries of unsustainable farming has caused some of the worst soil erosion on Earth; vast tracts of China's Loess Plateau are so badly eroded they can no longer support plant growth, and, each year, upwards of 1.5 billion tonnes of soil are washed from the Plateau into the Yellow River.

Soil erosion is a natural phenomenon and processes of soil formation to some extent naturally replenish losses of soil. The intensity of soil erosion also varies from place to place due to differences in the slope of the ground, texture of soil, vegetation cover, and climate. But one constant is that natural soil formation is painstakingly slow, and is overwhelmed by soil erosion from agricultural lands. To put this into context, rates of natural soil formation average around 0.1 millimetre a year, whereas rates of erosion in cultivated fields sometimes reach as much as 10 millimetres a year.[54] Because of this, fertile soil is being lost from farmland much faster than it is being replenished, leading to irreplaceable soil loss and declines in crop yield.

Another problem with intensive farming is that nutrients are exported from the soil in harvested crops, stripping the soil of its nutrient stock. Synthetic fertilizers can replenish macronutrients, like nitrogen and phosphorus, but crops also take up micronutrients such as magnesium, calcium, zinc, sulphur, and selenium, which are often not replenished. As a result, soils become progressively stripped of micronutrients, which eventually curtail crop yields. This is a particular problem where soils are very old and deeply weathered, and naturally nutrient-poor. Take India, one of the world's largest producers and consumers of synthetic fertilizers. To address increasing demands for food in the 1960s, agricultural reforms encouraged farmers to apply nitrogen fertilizers to their land. Initial benefits were dramatic: grain yields boomed, reaching record highs. But then grain yields started to drop, despite farmers still putting fertilizer to their land. Puzzled by this, scientists looked into the problem and found that, as well as depleting soil of organic matter, successive years of cropping had stripped them of their nutrient reserves, especially those not replenished by fertilizers, such as potassium and sulphur.[55] This created major nutrient imbalances, which curtailed not only crop yields, but also the ability of crops to respond to further dressings of nitrogen.

Heavy fertilizer use also comes with a serious cost to the environment. In many cases, farmers put far more nitrogen onto their land as fertilizers and manures than is needed by the crop, or it is simply applied at the wrong time, such as when the crop doesn't need it. The same applies to phosphorus; farmers frequently add more to their land, as manures and fertilizers, than is used by crops, progressively enriching the soil in phosphorus. This inefficient use of nitrogen and phosphorus is obviously wasteful for the farmer, but it also causes unintended problems for the environment. Large amounts of nitrogen and phosphorus not used by crops are washed out of soil into surrounding rivers and lakes, and even oceans, where they cause major problems for biodiversity and water quality. Also, in the case of nitrogen, under waterlogged or saturated soil conditions, specialist bacteria transform fertilizer nitrate into nitrous oxide, which is then released into the atmosphere, where it acts as a potent greenhouse gas. Nitrogen is also lost from soil as ammonia gas, which is very harmful to the environment. Ammonia is produced from the volatilization of urea, the preferred fertilizer of many farmers, and is a particular problem in wet, alkaline soils of high organic matter, which provide perfect conditions for rapid volatilization of urea.

The scale of environmental problems associated with heavy or inappropriate fertilizer use over the last fifty or so years is enormous. In many parts of the world, rivers, lakes, and groundwaters have become enriched with nitrate and phosphorus, creating problems for biodiversity and risks to human health. Emissions of nitrous oxide and ammonia from soil have soared, contributing significantly to global warming and pollution of the atmosphere; and many soils have become more acidic, a side effect of using too much ammonium-based fertilizer nitrogen. These environmental problems are common to most of the intensively farmed regions of the world, but they are especially acute in countries like China, where the use of synthetic fertilizers has soared in recent years, fuelled by government incentives. From the mid-1970s to 2005, grain yields increased by around 70 per cent, helping China to become self-sufficient, but over the same period the use of fertilizer nitrogen almost trebled and is used far in excess of crop needs.[56] As a result, soils of Chinese croplands have become more acidic,[57] crop yields have stalled, and nitrogen losses to the environment have doubled, causing serious environmental problems.[58]

SOIL DEGRADATION

If soil is worked too hard for too long, and not replenished, it will become exhausted and degrade. Whatever the cause, if a soil is stripped of its organic matter, biological diversity or physical integrity, or the plant cover that protects it from erosive forces or wind and water, its fertility will decline. The speed of soil degradation can be rapid: soils take literally thousands if not millions of years to form, but they can be destroyed in a matter of years. Soil degradation is now commonplace, with estimates suggesting that as much as 23 per cent of the Earth's usable land surface, which excludes mountains, deserts, and the poles, has been degraded to some extent, and each year, the world is robbed of an astonishing 5–7 million hectares of farmland because of soil degradation.[59] Even if these figures overestimate the extent of soil degradation, as argued by some, the worrying fact remains: soil degradation is widespread, causing often catastrophic declines in yield and, in many cases, a complete breakdown of food production systems.

I have already touched on soil erosion, which is the biggest cause of soil degradation, affecting around 2 billion hectares of farmland worldwide. As I have highlighted several times, soil erosion is a natural phenomenon. But if land is overgrazed, reducing plant cover and exposing soil to the erosive forces of wind and rain, or if it is over-cultivated, which robs the soil of its organic matter that binds soil particles together, or if unstable hill slopes are deforested and left bare, rapid erosion will set in. In fact, ever since humans began to cultivate the land, they have been plagued by soil erosion. The disappearance of the Norse from Greenland has been put down, in part, to excessive soil erosion caused by

overgrazing, which led to precipitous declines in grassland productivity and the collapse of Norse farming.[60] Catastrophic soil
erosion in Iceland, triggered by clearance of native woodland
and overgrazing, left much of the island barren and devoid of
soil.[61] The collapse of Easter Island has been attributed in part to
over-exploitation of resources, especially deforestation that caused
extreme soil erosion and diminishing supplies of food. And, in the
1930s, major dust storms swept across the American Great Plains,
ripping up parched, overworked soil, causing some farms to lose
most, if not all, of their topsoil. During this period, known as the
'Dust Bowl', soil dust played havoc with people's lives across
America, leaving many farms in ruin (Figure 7). Dust even fell as
far afield as New York City, where it reportedly fell dark one day at

FIGURE 7 During the 1930s, major dust storms swept across the American Great Plains,
ripping up parched, overworked soil, causing some farms to lose most, if not all, of
their topsoil.

noon from dust, forcing the chief of the newly formed Soil Conservation Service to proclaim:[62]

> when people along the seaboard of the eastern United States began to taste fresh soil from the plains 2000 miles away, many of them realized for the first time that somewhere something had gone wrong with the land.

Another soil problem that troubles growers is salinization, which is the process by which soluble salts build up in the surface soil. This is a particular problem in arid and semi-arid regions of the world, such as the Middle East, Australia, Southwestern United States, and southern Europe. In these regions, high rates of evaporation and rising water tables draw dissolved salts upwards from groundwater and deep in the soil to the surface, where they accumulate. Also, the lack of rain means that the salts are not washed away, which exacerbates salt build-up in topsoil. In coastal regions, large amounts of salt also come from the ocean, carried by winds and deposited on land in rainfall and dust, or from tsunami waves, which dump large amounts of salty seawater on land. This presents a host of problems for plants. It interferes with their normal metabolism and ability to uptake nutrients, and the high pH of salt-affected soils, which is usually around 8.5, hinders the growth of plants that can't cope with alkaline conditions. The problems don't stop there; the salty crust that forms on the soil surface impedes water movement through soil, and the salts interfere with soil structure, making the soil more vulnerable to erosion.

Many soils are naturally rich in salts, so-called saline soils. But a major conundrum for farmers in dry regions is that they have to irrigate their crops to make up the shortfall in water supply. This worsens soil salinization, the process of salt build-up in soil. An astonishing 250 million hectares of the land on Earth is irrigated, which amounts to 20 per cent of the world's arable land, producing about one-third of the world's food supply.[63] But the ability of this land to continue producing food is severally threatened, both by climate change, which is limiting water supplies, and by soil salinization, which is leaving land barren and unable to support crops. The scale of the effects of salinization on civilization can be enormous, as illustrated by the third millennium collapse of the once-prosperous Mesopotamian civilization, which occupied the vast alluvial plateau between the Tigris and Euphrates rivers. Several factors played their role in the collapse of Mesopotamia, including a shift towards a drier climate.[64] But soil salinization also played a major role, reducing the fertility of once-productive land. The soil of Mesopotamia was ideal for growing crops, being inherently fertile and flat. But the lack of rainfall meant that irrigation was needed to maintain crop yields for a booming population. However, the groundwater used to irrigate fields was rich in salts, and, combined with a baking sun and high rates of evaporation, this caused salts to rapidly build up in surface soil, effectively poisoning the soil. The way to combat this was to irrigate in moderation and leave land fallow; but growing pressure to produce food meant that these practices were bypassed, and soil fertility went into decline, as did food production and the population of Mesopotamia.

Today, salinization is a major problem in many arid and semi-arid parts of the world, and the area of land affected is growing fast. Something like 7 per cent of the word's land surface is affected by salinization, which is equivalent to an area ten times the size of France. In Mediterranean countries, for example, large areas of irrigated land are severally hampered by salinization, and much more is under risk. In Australia, where much of the land is naturally saline, farmers are so aware of the devastating effects of soil salinization that they refer to it as 'white death'. Literally millions of hectares of Australian farmland have been completely ruined by salinization, causing catastrophic damage to crops and farmers' livelihoods. In the United States, the government was so concerned about the troubles caused by soil salinization that they established a Soil Salinity Laboratory in 1937, to find solutions to farming salt-rich soils. And in China, soil salinization is spreading rapidly throughout the country, causing havoc for farmers. Unless solutions are found fast, the troubles of soil salinization will only get worse.

There are many other causes of soil degradation that I have not mentioned here, such as soil-sealing with concrete, asphalt, and bricks in towns and cities, and desertification, caused by poor land management and loss of plant cover, which reduces organic matter entering soil, and leaves them exposed to the vagaries of wind and rain. In other parts of the world, such as the Sahel region of Africa and the Qinhgai-Tibetan Plateau, China, catastrophic soil degradation has been put down to heavy livestock grazing (which has left land sparsely vegetated, compacted, and vulnerable to erosion) but also climate change (which has accelerated the speed of soil decay). And in many parts of sub-Saharan Africa, the traditional practice of leaving land fallow to restore soil fertility has been abandoned. As a result, soil degradation has intensified and crop yields have declined.[65] I could go on, but my message is simple: in many parts of the world, mismanagement of land is causing havoc for soils and their capacity to grow food, and, unless unchecked, these problems will only get worse under climate

67

change. Soil degradation played its role in the collapse of past civilizations and, given its strong link to poverty, it continues to hurt many people today.[66]

LIGHT AT THE END OF THE TUNNEL

I have so far talked mostly about soil problems caused by intensive farming and neglect of soil. But soil problems have also provoked government bodies and scientists to look for new, sustainable ways of managing soil. The 'Dust Bowl' provoked such a reaction. Awakened by the threat of soil erosion to the nation, US Congress directed the Secretary of Agriculture to establish the Soil Conservation Service, and following this, in 1937, President Roosevelt wrote to his state governors: 'The nation that destroys its soil destroys itself.'[67] Headed by Hugh Bennett, a pioneer in soil conservation, the Soil Conservation Service was charged with curtailing soil erosion and other soil problems, a role they still perform today.[68] At the same time, others were also looking for new approaches to soil management, centred on the fundamental role of soil organic matter. Sir Albert Howard, a pioneer of the organic farming movement, was one. In 1940 he published his famous book *An Agricultural Testament*,[69] in which he explained the fundamental importance of humus for soil fertility, arguing that humans should manage soils following nature's methods, paying attention to the balance between plant growth and decay. Drawing on years of experience of crop production in Indian smallholdings, he explained the critical importance of organic matter for soil structure, the supply of nutrients to plants, and the control of pests and diseases, and advocated the use of composting for restoring and maintaining the health of soil.

One of the most radical measures to emerge from the soil conservation movement of the mid-twentieth century was reduced, or no-tillage, farming. For centuries, farmers have ploughed their land to prepare it for crops, and the invention of the steel mouldboard plough by John Deere in the mid-nineteenth century paved the way for farmers to break the thick sod of native grassland and convert it to corn (Figure 8). In his controversial 1943 publication, *Plowman's Folly*, the agronomist Edward Faulkner dropped what was referred to as a 'bombshell' on the agricultural establishment, arguing that the mouldboard plough was responsible for widespread degradation of soil.[70] The basic idea of no-tillage farming is that the soil isn't ploughed, but rather seeds are dropped directly into a groove to minimize soil disturbance. Residues of the previous crop are also retained, so a large amount of the soil surface remains covered after planting. The benefits of this approach for

FIGURE 8 John Deere introduced the self-scouring steel plough, which allowed vast areas of native prairie to be ploughed and converted to cropland.

the soil fertility can be considerable: the residues and lack of ploughing protect the soil from erosion by wind and water, and promote the retention of water in soil and a rich soil life, which improves soil structure and nutrient supply to crops.[71]

Although it took a while for it to take off, vast areas of cropland in North and South America, and also Australia, are now converted to no-tillage, and as a consequence soil erosion has dropped considerably in some places.[72] But, the benefits are variable, and in some parts of the world it is not the best option: while no-tillage combined with residue retention is good for conserving soil water in dry regions of the world, in wet climates the lack of ploughing creates serious problems with weeds, resulting in heavy reliance on herbicides. This is a particular problem for smallholders in low-income countries where herbicides are not available or too costly, meaning that more labour is needed for hand hoeing. The cost of seeding machinery is also very high, preventing some farmers making the switch. Also, crop yields often fall during initial years of conversion to no-tillage, and extra fertilizer nitrogen is sometimes needed to counter this. Such concerns have led some to argue that the benefits of no-tillage are more limited than is often assumed, and that its expansion into some parts of the world should be done with caution.[73]

Another farming practice that is catching the eye of scientists is intercropping, which involves growing two or more crops alongside each other in the same field. This practice has been done for centuries, and is still widely practised by smallholder famers in many parts of the world, such as Africa, Latin America, and China. In Africa, maize is commonly grown along with cowpeas or beans; in Latin America, maize is grown with beans, potatoes, and squash; and in China, maize is grown alongside soya and fava beans. Intercropping was also commonplace in the US and Europe until the advent of the Green Revolution, when high-yielding monocultures became the norm.

The general idea behind intercropping is that the different crops complement each other, leading to more stable, bigger yields, and less risk of crop failure. The benefits can be far-ranging. Growing nitrogen-fixing legumes, such as peas and beans, alongside maize or wheat not only boosts soil nitrogen levels, but also smothers weeds and covers the soil, protecting it from erosion. In Africa, a common practice is to grow nitrogen-fixing, 'fertilizer trees', along with maize, sorghum, and millet, which can boost yields of these crops, whilst also improving the soil and providing firewood and fodder for livestock. Intercropping can also improve water use by crops, with the roots of different crops exploiting water from different zones of the soil, and it can discourage pests and diseases, with one crop acting as a deterrent to the pests and diseases of another, or as a 'trap' for pests that would otherwise attack the main crop. A well-known example of using crops as a deterrent is the growing of onions alongside carrots, which mask the smell of carrots to its pests. Planting grasses among maize or sorghum, or around the edges of fields, is also commonly used in Africa as a 'trap' for insect pests of the principal crop.

Scientists are now taking a fresh look at intercropping to see whether it can be tuned using new scientific knowledge to make farming more sustainable and to improve soil.[74] For example, it is now possible to select crops that release specific enzymes or organic compounds into the soil that mobilize phosphorus, making it available to themselves, but also non-phosphorus-mobilizing crops that grow alongside them, thereby boosting overall yields. Chickpea is one such crop that can mobilize phosphorus in soil, increasing its uptake by wheat, and fava beans can also make phosphorus more available to maize. In a similar way, some crops release organic compounds from their roots into soil that make micronutrients—such as iron, zinc, and manganese—available to companion crops, again boosting yields. This kind of scientific knowledge opens the door to the idea of selecting different crop varieties, or genotypes, that perform better together than when grown alone, or which together promote the growth of micro-organisms that mobilize nutrients in soil. It also opens the door to the breeding of crops with special characteristics, or traits, that benefit the growth of neighbouring crops, whilst also improving soil. This might be far off, and the considerable challenge of mechanically harvesting mixed crops still remains. But selecting and breeding crops to grow better together whilst also improving the soil could be an effective way to solve many problems of modern farming and its negative effect on soil.

Lessons on building soil fertility also come from deep in the Amazon. Soils of Amazonia are typically infertile, highly wea-thered, and challenging for growing crops. But, as I touched on in Chapter 1, throughout Amazonia are found patches, ranging from less than a hectare to several square kilometres, of very fertile, dark soil, called terra preta soils. These soils, which cover as much as a tenth of Amazonia,[75] are rich in organic matter and nutrients, and support much higher crop yields than surrounding soils. Moreover, they have been cultivated for centuries and remained fertile to this day.[76] Puzzled by their existence, scien-tists have carried out detailed investigations into their origin, and it turns out that they are of human origin; they were mostly formed between 2500 and 2000 BP by pre-Columbian native populations who added tremendous amounts of charcoal, organic wastes, excrement, bones of mammals and fish, and even pottery to land.[77]

So what has made these soils so fertile? This is a question that has puzzled many scientists and archaeologists, who have done various tests to unravel the cause of their remarkable fertility. From what they have found, two main factors have contributed to their formation: many years of charcoal addition have incorporated nutrients into soil, especially potassium and phosphorus, and increased nutrient retention, whereas years of adding organic wastes, excrement, bones, and pottery, built up stocks of nitrogen, phosphorus, and other nutrients such as calcium.[78] The result is an anthropogenic soil with an unusually high organic matter and capacity to retain nutrients, and an ability to supply a wide range of nutrients to crops in a slow and efficient way.

It has been suggested that these ancient terra preta soils could hold clues for managing tropical soils today.[79] The traditional farming system in the tropics is slash and burn, whereby farmers slash the natural vegetation and burn it, making way for crops. Under this system, soil fertility declines rapidly, so after a few years farmers would leave land fallow for it to regenerate. Increasing population pressure, however, has resulted in more tropical forest being converted to cropland, and more land being cultivated without fallow, resulting in major problems for soil fertility. While organic manures and mulches can help to keep these soil problems in check, the warm, wet conditions in the tropics mean that soil amendments of easily degradable organic matter are rapidly broken down, and its benefits are short-lived. To overcome this, it has been suggested that soil amendments of easily degradable organic matter could be replaced by charcoal, effectively simulating the formation of highly-fertile terra preta soils.[80] This could be done by replacing the burning in slash and burn systems by charring slashed residues, or charring different organic wastes, such as rice husks or urban waste, to produce biochar, which can then be added to soil. Field trials in the tropics show that charcoal additions can increase crop yields, but, as with slash and burn, the benefits don't appear to persist. However, by combining charcoal additions to soil with other organic wastes, such as manures, it might be possible to replicate some of the longer-term benefits for soil fertility seen in terra preta soils.

These are just a few of the ways in which scientists are looking to improve the way growers manage soils and make farming systems more sustainable and less damaging to the environment. There are of course many others, such as bringing back mixed farming to ensure adequate supply of manures; the use of composts and micronutrient amendments to correct soil nutrient imbalances; and the breeding of crops to improve their ability to acquire valuable nutrients and work with soil organisms to better do this. As I discussed in Chapter 2, there is also the possibility of promoting the biodiversity of soil to make nutrient cycling more efficient and less leaky. But central to them all is the principle that a healthy crop requires a healthy soil, and hence sustainable food production needs to take account of the natural processes that build and maintain the fertility of soil. Intensive farming has clearly lost sight of these natural processes, and the consequences of this for the grower and environment are plain to see.

TERROIR

A group of growers that have an especially intimate relationship with the soil are the makers of wine. This deep association between soil and wine goes far back in time, to the Middle Ages, when, legend has it, Cistercian monks in Burgundy, France, marked out where to grow their wine by tasting the soil. Also from their traditions emerged the notion of terroir: the aspects of climate, geology, soil, and human culture that create unique characteristics in wine.[81] The notion of terroir is a matter of considerable debate,[82] but at its heart is the idea that the character of a wine depends on its place of origin. But which factors are most important? The way that grapes are grown and the winemaking process obviously plays a role, as does climate, which determines not just where wine is produced, but also the variety of grape grown. Nuances in weather, such as frost frequency, rainfall, and drought, also have consider-able impact, as does the slope and orientation of the land, which affects the amount of sunshine reaching the vines.[83]

Many winemakers also believe that the soil is a major part of terroir, and that the character of wine of many regions is intim-ately linked to soil. In Bordeaux, for example, wine quality is often put down to the deep, gravel- and sand-rich soils of the region, which promote rapid drainage and deep root penetration, helping vines to survive periods of heavy rain and drought. Similarly, the best wines of Napa Valley in California are grown on well-drained, gravel-rich alluvial soils, and in South Australia, the success of vineyards in the Coonawarra region is put down to a narrow strip of terra rossa soil, or red earth, which creates perfect drainage conditions for growing grapes. The best wines of Burgundy are produced from soils rich in clay and pebbles, which create a perfect balance of good drainage and water retention; in Spain, the character of Priorat wines, which are among the world's most expensive, is defined by the unique, freely drained, nutrient-poor *llicorella* soils of the region.

Although these examples suggest a strong historic association between soil and wine, scientific evidence for this link is sparse. Of course, vines rely on soil for nourishment, so differences in soil properties, such as soil depth and texture, that affect how much water or nutrients are available to vines, will affect the quality of wine. Also, scientific studies show that soil water availability is of paramount importance for grape quality.[84] So it follows that soil properties such as texture, which controls water flow through soil and the amount that is retained, and soil depth, which dictates how deep roots can grow to get water, will play an important role. But the question is not whether soil variation affects wine quality, which it clearly does; it is more how important soil is for terroir relative to other factors, such as climate and the methods used to produce the wine.

To learn more, I decided to visit Burgundy, a region that produces some of the finest wines in France, and has a deep connection with the concept of terroir. This area has an enormous variety of soils, and is parcelled up along limestone escarpments into a patchwork of small vineyards, or *climats*, which are thought to represent the very expression of terroir. During my visit, I heard a variety of contrasting views on the importance of soil for terroir. I heard from soil scientists, who told me about the sheer scale of soil variation in the region: soils varying greatly in their depth, stoniness, and clay content across very short distances, affecting soil drainage and the extent to which grapes suffer water stress during periods of drought, which ultimately influences the quality of wine. I heard from others that the soil didn't actually matter so much for terroir, and that the way that the grapes are grown and the quality of the winemaker play a much bigger role. And as another twist to the tale, I heard how in some parts of Burgundy, the soil has been washed away from vineyards by centuries of erosion, and was replaced during the Middle Ages by workers who would regularly carry large baskets of soil from the bottom of slopes to the tops, or even from entirely different places, to replenish the eroded soil (Figure 9).[85] This discovery left me questioning the importance of soil for terroir, given that soils of these exclusive vineyards are not so distinctive after all.

The best test of the contribution of soil to terroir is to taste the wine. For this, I visited Pierre Cornu, the proprietor of *Domaine*

FIGURE 9 Labourers at the end of the Middle Ages transporting soil in vineyards to counter soil erosion.

Edmond Cornu et Fils, a wine-growing company in Ladoix, Côte de Beaune, established in 1875. Pierre is a strong believer that soil expresses its nature in his wines; his wines all come from nearby *climats* with different soils, but similar climate, and his grapes are all treated the same; they are handpicked, fermented in steel tanks, and aged in oak barrels for similar amounts of time. Differences in his wines must therefore be mostly related to soil. Pierre described how the taste of each of his wines reflected the soil: the more robust, powerful wines, which he likened to a rugby player, came from heavier, clay soils, whereas the more elegant wines came from deeper, gravelly, freely drained soils. He didn't have a scientific explanation for these differences, but it was clear that the soil expresses itself in his wine.

I only spent a short time in Burgundy, but from my visit it became abundantly clear that many factors contribute to the terroir. Even the species of oak used for the wine barrels leaves a distinct chemical signature on wine,[86] as does the environmental conditions under which the wine is grown.[87] But as with any crop, the soil nourishes the vines with nutrients and water so any difference in soil, either natural or man-made, will leave its mark on the quality of wine. The extent that soil expresses itself in wine relative to other factors of course varies from place to place, and winemaker to winemaker. But tasting the wines of Burgundy convinced me that, in some places at least, the soil does contribute to the terroir.

Soils are remarkably resilient, but they need to be cared for. At the heart of a fertile soil are natural processes of organic matter recycling, the efficient supply of nutrients to plants at critical times of growth, which is driven by the rich diversity of soil life, and the maintenance of soil structure, which allows roots to exploit the soil for water and nutrients, and for water and gases to move freely though the soil profile. History tells us that neglect of these natural processes leads to rapid declines in soil health and crop yields, creating problems for society at many levels, for the individual grower and for nations alike. History also tells us that these problems can be catastrophic, such as in Mesopotamia and during the Dust Bowl in the US. But soil problems can also be resolved; by treating soil as an investment and integral part of the grower's ecosystem, taking care to ensure that it is replenished and that the natural processes that gave it fertility in the first place are nurtured, its fertility can be maintained. Agricultural technology offers ways of refining the way we manage soil, for example by combining crop breeding with good soil management to encourage crops to use resources more efficiently. But any amount of crop breeding or agricultural technology, however sophisticated, means little if soil is neglected and its fertility has declined. Whatever the crop, whether it is a cereal, grass, or grape, it needs a fertile soil to grow, and the consequences of neglecting soil fertility at the expense of short-term yield are plain to see.

4

Soil and the City

To forget how to dig the earth and to tend the soil is to forget
ourselves.   *Mahatma Gandhi*

I have spent most of my living and working life in the countryside,
surrounded by open fields, woodlands and hills, and in close
contact with the soil. I recently changed my job and moved to
the University of Manchester, which is in the centre of one of the
largest cities in England. Because of this move my contact with soil
is much less; in fact, as I walk each morning to my office, there is
hardly a handful of soil to be seen.

But is this really true of the whole city? Concrete, asphalt, and
bricks certainly seal much of the ground in Manchester, as in most
cities and towns. But soil is in abundance: it lies beneath the many
small gardens, flower beds, road and railway verges, parks, sports
grounds, school playing fields, and allotments of the city. In fact, it
has been estimated that almost a quarter of the land in English
cities is covered by gardens, and in the United States, lawns cover
three times as much area as does corn. As I write, I am on a train
leaving central London from Waterloo Station, and despite the

overwhelming dominance of concrete and bricks, I can see scattered around many small gardens, trees, flowerpots and window boxes, overgrown verges on the railway line, small parks and playing fields for children, football pitches, grassy plots and flower beds alongside roadways and pavements, and small green spaces with growing shrubs outside office blocks and apartments. The city is surprisingly green and beneath this green is soil.

Throughout the world, more and more people are moving to cities: in 1800 only 2 per cent of the world's population was urbanized, whereas now more than half of the global human population live in towns and cities, and this number grows by about 180,000 people every day.[88] This expansion has been especially rapid in recent years. In 1950 there was only one city in the world with a population of more than 10 million, New York City, whereas now there are upwards of twenty, and most of them are in the developing world. In my own country, England, urban areas cover over a tenth of the land surface,[89] and across the whole of Europe around 1 million hectares of land were taken over for urban use during the decade 1990–2000.[90] The picture is no different in other parts of the world: globally, the amount of urban land surface is predicted to double within the next twenty years, and perhaps even triple in developing countries. Such rates of expansion will cause considerable loss of soil due to the sealing of land, and it will also place a major strain on unsealed soils and their ability to perform important functions for the people who live there (Figure 10).

ASPHALT, BRICKS, AND CONCRETE

The most catastrophic impact of urban expansion on soil is through sealing, or the covering of soil with impermeable materials

FIGURE 10 The expansion of cities is consuming vast amount of prime agricultural land and soil. Urban expansion was greatest during the second half of the twentieth century, and globally it is predicted that the urban land surface will double within the next twenty years, and perhaps even triple in developing countries.

such as asphalt, bricks, and concrete. The scale of soil sealing is such that many consider it to be the biggest threat to soils globally because it takes up fertile agricultural land and often causes irreversible loss of the natural functions of soil. As I have already mentioned, rates of urban expansion are accelerating and with this comes the sealing of soil. Across Europe, for example, around 500 square kilometres of land are sealed by impermeable material every year, which is an area roughly half the size of the city of Berlin.[91] And in Germany alone about 27 hectares of land are sealed every day, which is the size of around thirty-five football pitches.[92]

Sealing effectively suffocates the soil. It prevent plants from being able to grow and abruptly ends the many functions that soils perform, such as the recycling and storage of water or nutrients, and the exchange of gases between the land and atmosphere. The consequences of this can be dramatic. The sealing of soil with paving and roads makes the land surface virtually impermeable to water, which increases surface run-off following rainfall events because water can no longer seep into the soil. This can place considerable pressure on sewerage systems, and often causes severe flooding in areas where run-off water collects. Few studies have been done to record such problems, but of note was one done in a suburban area in Leeds, a large city in northern England. By studying aerial photographs from 1971 to 2004, it was found that there had been a 13-per-cent increase in impervious surfaces over that period, which was mostly down to the paving of front gardens to provide parking for cars. Moreover, this caused a striking 12-per-cent increase in annual surface run-off, placing considerable pressure on the city's drainage system.[93] Similar patterns were also found in Leipzig in Germany, with surface run-off more than doubling between 1940 and 2003 due to an increase in the sealing of the city's soil.[94]

Soil sealing can also affect heat exchange in cities. This is because asphalt and concrete have a lower albedo, or reflectivity to sunlight, than unsealed soil with vegetation. Because of this, more radiation is intercepted by sealed surfaces, which makes them and the surrounding air much warmer, ramping up the so-called urban heat island. In London, for example, daytime temperatures can be up to 9 °C greater than in surrounding rural areas, an effect that is mostly put down to the abundance of heat-absorbing surfaces.[95] Soil doesn't directly cause this warming effect. But urban planners are beginning to realize that unsealed soils, and plants that grow on them, can help to dampen the urban heat island. Trees, for example, consume large amounts of energy to drive the uptake of water from soil and its evapotranspiration, and in doing so they can lower summer temperatures in cities. They also produce shade, which helps keep summer warming down and reduces the need for air conditioning. Another profound impact of soil sealing is that it drastically reduces the exchange of gases between the soil and atmosphere. As I will discuss in Chapter 6, soils impact on climate because they store large amounts of carbon, which comes from plants that absorb carbon dioxide from the atmosphere; sealing puts a stop to this.

Soil sealing has ramped up in recent years as urban areas expand, but it is not a recent phenomenon; cities have been growing for centuries. The city of Manchester, where I work, was little more than a small market town at the end of the eighteenth century. But within just over a century, it had grown at an astonishing rate to become the world's first industrial metropolis and one of the most populous cities in the world. This expansion was brought about by the Industrial Revolution, which led to massive areas of fertile soil being consumed by factories, roads, and houses as the city sprawled into surrounding farmland. Similar engulfing of soil by growing cities happened in many parts of the world. But what has changed in recent years is, first, that the rate of urbanization has surged, and second, it is dawning on city planners that sealing can severely impact the urban environment and, conversely, that unsealed soils play a crucial role. Because of this, there are now calls for urban expansion to be more soil-friendly by restricting sealing to poor quality soils, unsealing sealed soils that are no longer in use, and offsetting soil sealing with unsealing or greening measures in other parts of a city, thereby reducing the total area of sealed soil.

THE HIDDEN SOILS OF THE CITY

Although soil sealing is happening at a rapid rate, much soil remains unsealed, often in the most unexpected places. The most obvious place to find soil is in the many gardens, parks, and allotments that are scattered throughout the city. But soil also plays an important role in cemeteries and churchyards, sports pitches, urban nature reserves, and even rooftop gardens that are becoming increasingly popular in urban areas. The amounts of greenspace underlain by soil can be remarkable. A recent survey in Britain[96] discovered that the amount of green space and domestic gardens in London is about 60 per cent. In Birmingham, the second largest city in England, this figure is even higher at 63 per cent, and in Newcastle upon Tyne, in north-east England, it reaches an astonishing 71 per cent. Liverpool is lower at 38 per cent, but even this represents a surprisingly large amount of green space and unsealed soil.

Urban soils come in many forms, but what most distinguishes them is that they are drastically affected by human activity, to the extent that soil taxonomists classify most urban soils as 'anthropogenic'. Take New York, for example, one of the largest cities in the world. Here, soil scientists recently completed an extensive soil survey—the New York City Reconnaissance Soil Survey—to describe the city's soils and map soil patterns across the New York City territory.[97] This pioneering survey, which was a collaborative effort between the United States Department of Agriculture (USDA), the New York City Soil and Water Conservation District, and soil scientists at Cornell University, came to life because of a recognition that the city's soils were a crucial element of its environmental quality and that a soil map would be invaluable in deciding how to best use the city's land.

The survey revealed an astonishing diversity of soils, some of which that had never been described before. A total of forty-six different types of soils were found, and while some were formed naturally over thousands of years on the glacial deposits that underlay the city, most of the soil types discovered were relatively young and formed on human constructed material. The aptly named Bigapple soil, for example, is formed of dredged sand along New York City bay, and is a deep, freely drained soil of a loamy sand texture. Freshkill soils are formed in human refuse over landfills, and are well drained, deep, and of a silty or sandy loam texture. Inwood soils form on demolished construction rubble, and are very deep, well drained, and of a silty loam texture. And Fishkill soils are formed in incinerator fly ash, and are very deep, but poorly drained. These are just a few of the soil types that the surveyors found in New York City and which are likely found in other urban areas around the world.

This soil survey is one of the few that have been done in cities, which seems an incredible oversight given the many crucial roles that soils play in the urban zone. Not only do they absorb water, acting as a buffer against flood risk after heavy rain, but they also absorb heat, keeping in check the urban heat island. Urban soils also retain nutrients, preventing them being lost in run-off and drainage waters, and they help to degrade pollutants of the city. They can also store vast, but often unaccounted-for, amounts of carbon, thereby helping to mitigate climate change. In fact, amounts of carbon in urban soils, beneath buildings, pavements and residential driveways, and in urban green spaces, such as parks and gardens, can be much greater than in agricultural land.[98]

Soils play their biggest role in the myriad gardens, parks, and allotments that are scattered throughout the city. Here, soil supports the plants that adorn our gardens and city parks, it underlies our garden lawns and parkland, and it provides nutrients and water for the food that city dwellers grow. It has been estimated that around 800 million city people worldwide grow their own food in one way or another, often by necessity. City people produce food in many ways. Some grow vegetables in garden pots on their windowsills, balconies, patios, and rooftops. Others grow vegetables in small garden plots or in greenhouses in their backyards, and some cultivate the soil in large urban farms. But in many parts of the world, the allotment, or community garden, is where most people produce their own food, and is where humans have close contact with soil (Figure 11). This is especially the case in Europe, where allotments are commonplace

FIGURE 11 Demand for allotments, such as these close to Old Trafford Football Ground, Manchester, is booming, as city people turn to the soil to grow their own food.

in most towns and cities. But they have now spread to many parts of the world.

The use of allotments reached its peak during the First and Second World Wars, as civilians in many countries were encouraged by their governments to grow their own food to overcome food shortages. The number of allotments in the UK, for example, doubled during the Second World War to reach a staggering 1.6 million because of the 'Dig for Victory' campaign. And in the US, the 'Victory Garden' was a mainstay for food production for civilians threatened by food shortages during the war. Likewise in Germany many took to their spades and allotments to produce food; I recall visiting Jena and being taken aback by the number of allotments that surrounded the city, which I was told mostly originated from the Second World War.

Although interest in allotments dipped after the Second World War, they have exploded in popularity in recent years as more and more city people turn to the soil to grow their own. In the UK, demand for allotments currently outstrips supply, with an estimated 90,000 people sitting on waiting lists for vacant plots. Demand for allotments is also increasing in other countries: in Germany around 4 million people flock to their allotment gardens in the outskirts of cities to turn the soil, and in the USA and Canada there has been a boom in the number of urban community gardens in recent years, where groups of people congregate to grow their own food. Many things have spurred this resurgence in allotments, the most obvious being greater awareness of the health benefits of growing your own food. But there are other benefits, such as getting away from the hustle and bustle of city life, improved physical fitness, cheaper access to healthy food, and perhaps most importantly, knowing the provenance of your food; when it comes to food, there is nothing more reassuring and enjoyable than eating your own.

You might expect that soils of the city are not as fertile as those of farms in rural areas that produce most of the world's food. A study by a team of British scientists showed that this wasn't the case.[99] Their study was based in Leicester, a relatively small city in the East Midlands of England, and the team found soils taken from allotments contained much greater amounts of organic carbon and nitrogen than did soils taken from surrounding agricultural fields under intensive arable production. They also found that soils from allotments were less compacted than agricultural soils, or in other words they had more pore space and hence were better able to allow free gas exchange and drain water. Given that these measures all indicate good soil fertility, they concluded that, at least in this area of England, soils of allotments are of considerably better quality, and most likely support higher crop yields, than are those of surrounding arable fields.

Why should this be the case? The main reason for the higher soil quality in allotments is that most of their owners use substantial amounts of manures and composts, which boost organic matter and nutrient content of soil. Almost all of the allotment owners of the Leicester study added compost to their soils, made from both allotment debris and household waste. On top of this, most of the owners added manures to their plots, which will also have boosted organic matter and nutrient content of their soil. By contrast, the arable soils that surround the city will have received no manure or compost, and the only input of organic matter would be the remains of the crop, the stubble and roots that are ploughed into the soil each year following harvest. On top of this, year after year of tillage and heavy fertilizer use will have depleted these soils of organic matter, leading to progressive decline in their fertility.

It would seem therefore that urban soils can maintain extremely high levels of fertility, much higher than commonly found in surrounding areas where land is intensively farmed. Of course that was just one study, in one small area of England, and it might not reflect what happens in the rest of the world, or for that matter in parts of England. But as the authors of the study noted, records of incredible yields being produced from allotments during war years in Europe and the USA, combined with results from recent UK trials which show fruit and vegetable yields from allotments to be four to eleven times greater than those of major agricultural crops, suggest this could be the norm.[100]

While all of the above points to the merits of allotments relative to agricultural lands for the fertility of soil, large-scale production of food in urban allotments has a long way to go before it can be considered sustainable. As the authors of the Leicester study noted, many of the inputs that bring fertility to allotment soils, such as manures and vegetable wastes, come from the agriculture and fisheries industry. Also, many allotment owners rely on chemical fertilizers to maintain their yields, which come from the petrochemical industry. Having said this, what is clear is that there is much potential to boost food production in urban areas, as was done during the wars, and also to make it more sustainable, for instance through large-scale, citywide composting of household wastes. But what is also abundantly clear is that the fertility of urban soils is not a limiting factor.

THE SPORTS PITCH

A lasting memory of watching English football as a child is that the pitches were often quagmires of mud, especially during winter, and players left the pitch caked in mud (Figure 12). Things have certainly changed and now football pitches of the top clubs are in pristine condition and players rarely leave the pitch with the stain of mud. There are several reasons for this change, but high on the list is the fact that the financial stakes of football are so high that maintaining an excellent pitch in all weathers is crucial for the success of the team. On top of this, knowledge and expertise in the science of turf culture has advanced greatly, and the groundsman is considered key to the team's success.

FIGURE 12 Two Preston players leaving the quagmire of a football pitch at half-time during a match against Charlton at The Valley in 1937.

Sports have always played a major role in cities. Take Manchester: 75,000 people flock to Old Trafford to watch every home game of Manchester United, and around 47,000 people pour into Manchester City's ground to watch them play at home. Just a short journey to the west, around 45,000 people flood into Anfield to watch Liverpool play, and across all major cities of the world, such as London, Barcelona, Madrid, and Milan, legions of people go to football grounds every weekend to watch their team. This is just one sport, but there are many others that go on every day in cities, such as cricket, tennis, rugby, golf, and baseball, which bring enormous pleasure to people; for all these sports, the management of turf and its underlying soil plays a significant role.

Turf management is not straightforward. Different sports require different turf characteristics, and what makes for a good turf in one part of the world, will be different to what is best in another. For football, the main characteristics of a good pitch, as recommended by FIFA, are that it needs to be free of dips and hollows, well drained with good grass cover of desirable species, firm and stable to give good playing characteristics, and appropriately marked with good visual appearance. But the methods used to construct pitches vary enormously, depending on the quality of the pitch required and weather conditions. For a stadium hosting professional or international football, for example, the pitch needs to be of the highest standard in all weathers, such as freezing, drought, or heavy rain, to take out the risk of poor playing conditions or even cancellation. To achieve this, pitches are meticulously constructed with gravel subsoil and sand-dominated upper layers, combined with sophisticated drainage and irrigation systems to create an optimal rooting zone for grass growth. Depending on the prevailing weather, this might be accompanied by underground heating or even artificial lighting where shading is an issue, which is often the case in large stadiums. Some go even further: at Old Trafford, home of Manchester United Football Club, the head groundsman uses a battery of soil moisture sensors and pitch cameras to control irrigation, and under-soil heating to ensure the pitch is in its best form.

Unlike major football stadiums, standard pitches, such as those of amateur football clubs or schools, or those found in city parks, have to make the best of the natural soil that underlays the pitch. Ground staff can do their best to improve playing conditions through levelling, soil drainage, mowing, adding fertilizer, and sowing with appropriate grass species. But certain soil-related problems are hard to overcome. For example, soils are notoriously variable and even subtle differences in the texture or acidity of soil across a pitch will cause patchiness in grass growth and therefore playing conditions. Even when soil conditions are uniform across a pitch, differences in soil type and weather have a major impact on playing conditions: pitches on clayey soils are more likely to become a quagmire during wet winters, and in colder, wetter climes, pitches are often susceptible to a build-up in decaying organic matter on the soil surface, forming a dense thatch of dead roots and shoot matter. Not only does this thatch hamper drainage, making the ground even more sodden, but it also affects the firmness of the pitch and the way that the ball rolls and rebounds on the surface.

Although mostly on the outskirts of towns and cities, another sport that demands pristine soil is golf. Golf courses are subject to some of the same soil problems as are football pitches and other turf-based sports, but they differ in that they cover much larger areas of land. Indeed, the amount of land taken up by golf courses in and around urban areas can be considerable: a recent survey in Britain found that 2.8 per cent of the land area of Merseyside, which encompasses the city of Liverpool, is given over to golf, and close behind at 2.7 per cent is the West Midlands, which includes the major cities of Birmingham, Coventry, and Wolverhampton.[101] In the US, it has been estimated that golf courses cover over 2 million acres of land, which amounts to about 3,500 square miles,[102] and in Bermuda, the most golfed place on Earth, just over 8 per cent of its land is given to golf.[103]

Considerable effort is put into ensuring optimal soil conditions on golf courses, and this couldn't be more so than on the putting green. Here, great attention is focused on the design and construction of the soil, because even subtle differences in soil properties can have a considerable effect on play. For this reason, soil profiles of putting greens are mostly constructed using soil mixes that best balance the need for uniform grass growth, rapid drainage, resistance to compaction, and an ability to retain water between irrigation cycles or rain events. As on football pitches, sand-dominated soils are the norm because of their effective drainage. But sand is not so good at holding water or nutrients, so to counter this, soil amendments, such as peat and expanding clays, are often added to perform these roles. Indeed, most putting greens are constructed following strict guidelines of organizations, such as the United States Golf Association, who provide detailed guidance on the construction of golfing soils, from the depth and spacing of drainage channels right down to the particle size of the gravel and sand used in the soil mix.

Despite this attention to detail, golf course soils present their owners with many challenges. One of the biggest headaches is hydrophobic or water-repellent soil, which forms localized dry spots on the putting green, which interfere with irrigation systems, often rendering them ineffective. These localized dry spots also cause severe wilting or death of turf grass, creating an irregular green surface and big problems for the golf course superintendent. Hydrophobicity is a natural process whereby soil mineral particles, and especially sand, become coated in organic substances sourced from the decay of dead plant remains and organic amendments, such as peat. These coatings repel water and, as a result, water beads up on the soil surface instead of penetrating through soil, thereby drying out the soil. Golf courses are especially vulnerable because they use sand-rich soil mixes and organic amendments, such as peat, which decay to form water-repellent coatings on the sand particles. Treatment involves the use of wetting agents, or surfactants, which are able to link hydrophobic soil with water, thereby increasing water movement through soil, although their effectiveness depends on many factors, including the type of soil, climate, and the chemistry of the wetting agent. There are many of these products on the market and choosing the right one can be a nightmare for the golf course superintendent.

Another cause of localized dry spots on golf courses, and other sports turfs, are fungi that live in soil: the fairy ring fungi. These are a specialist group of fungi that form circular or semicircular rings of mushrooms on the turf. These rings are not only unsightly, but they also create irregular turf conditions, which isn't conducive to good play. Treatment is problematic, and ranges from the use of fungicides to complete soil replacement: infected turf and underlying soil is removed and replaced by sterilized topsoil before the turf is reseeded. Another soil problem is the earthworm. Most hail the earthworm as a bastion of soil fertility, but on golf courses, earthworms bring problems: the earthworm cast. Earthworms consume vast amounts of organic matter along with mineral material, which are churned up in their guts. Some earthworm species deposit this material in soil, but others void it on the soil surface as casts, which create a maintenance nightmare: they compromise playability, cause trouble for aesthetics, attract predators such as moles, and they damage mowing equipment, blunting cutting blades. For these reasons, earthworms are widely considered as one of the main problems for golf courses and are therefore kept in check.

I have touched on just a few sports that are played in and around urban areas of the world, but many of the soil challenges, such as providing effective drainage, nutrient cycling, and soil uniformity to maintain turf conditions, are common to various other sports: management of soil hydrophobicity and fairy rings is critical for maintaining an effective bowling green or tennis court; a well-structured soil with a healthy sward is key to the game of rugby; and the art of cricket groundsmanship is to balance soil moisture with pitch playability. In other words, management of soil conditions and sport go hand in hand.

CONTAMINATED SOIL

Urban soils bring many benefits to humans, supporting rich crop yields, pristine sports fields, and beautiful parks and gardens. But they can also be contaminated by past or current industrial activity, or by simply being close to busy roads. These contaminants can be far-ranging depending on their source, and include heavy metals, such as lead, cadmium and mercury, organic chemicals, and even asbestos. Contaminated soils thereby pose a potential health risk to those who use them, either for building on, playing sports, or growing crops and gardening. Gardeners in particular can be exposed to contaminated soil by inadvertently ingesting or inhaling soil particles, or through eating vegetables grown in contaminated soil. Heavy metals such as lead, cadmium, and arsenic can accumulate in vegetables grown in contaminated soil, posing a potential threat to those who consume them. Despite this, a recent survey done in Baltimore, USA, suggests that most gardeners do not even know if their soils are contaminated and they are unaware of the health risks posed and how to manage them.[104]

An obvious question is: how contaminated are urban soils? Many researchers in many cities have tackled this question across the world, and a key message is that soil contamination is highly variable. It depends on many factors, such as proximity to current pollutant sources, including busy roads and incinerators, the industrial history of a site, and the properties of the soil itself. Nevertheless, what studies do consistently show is that soil contamination in the world's cities is widespread and comes in many forms. A recent analysis of soils from major cities in China, for example, revealed that heavy metal contamination is widespread, mostly from traffic and industrial sources.[105] And scientists working in Baltimore, the largest city in Maryland, USA, revealed that levels of lead, cadmium, copper, nickel, and zinc, were much higher in inner city gardens than in nearby agricultural soils.[106] Topsoil concentrations of a bunch of heavy metals, including lead, zinc, and mercury, were found to be greater in the city of Palermo, being mostly sourced from traffic, than in natural soils of surrounding areas of Sicily.[107] I could go on, but the message is clear: contamination of urban soils with heavy metals is widespread.

Heavy metals are not the only contaminants of city soils. A host of industrial activities, such as power generation, metal processing, chemical manufacture, and incineration of municipal and chemical wastes, produce organic pollutants called dioxins, which are highly toxic compounds that pose a significant health risk to humans. In many countries emissions of dioxins have dropped in recent years because of government controls. Nevertheless, they can cause a problem for many years because they bind tightly to soil organic matter and they are very tough for microbes to break down. Because of this, dioxins can linger in soils for many years, representing a lasting fingerprint of decades of past industrial activity. In fact, as with heavy metals, dioxin contamination of soils of the world's cities and industrial areas is widespread, with national surveys showing that soil concentrations are often several orders of magnitude higher than in rural soils in surrounding areas.[108]

Soil contamination causes a host of problems for humans. Take the boom in urban gardening that is happening across the globe. Most urban growers are simply unaware of the contaminants that lurk in their soil, and even if they were aware, they probably wouldn't know how to go about getting their soil tested for them. Even if a gardener did know where to go for soil tests, it would cost them a small fortune to test for the full range of pollutants that might be lingering in their soil. As a result, those that do have concerns, and the funds to pay, often limit tests to just a few soil samples and to areas that are most likely contaminated, such as close to busy roads or a factory. They might also restrict analysis to a single pollutant, such as lead, in the hope that it might act as alarm bell for other contaminants in soil. An alternative approach is to look into the history of a site for clues as to which pollutants might be there, or, as many do, simply assume the soil is contaminated and take measures to reduce exposure to pollutants that might be there.

So what can a gardener do to reduce the risk of exposure to pollutants? In extreme cases, the best solution is to completely remove the contaminated soil, seal the underlying ground with a water-permeable fabric barrier to block off pollutants deeper in the ground, and then cover with clean topsoil. Although effective, this is a costly solution, and is reserved for only the most extreme cases. An alternative approach is to build raised beds with clean soil, or to mix compost or manure into soil, which not only boosts soil fertility, but also dilutes the pollutant and binds them to soil particles, rendering them less available to plants. Other measures might be to build gardens away from roads or railways, and to build hedges or fences to reduce windblown contaminants coming into gardens. Given that the main route of exposure to pollutants is accidental ingestion of soil, either through inhaling dust or ingesting contaminated soil particles from dirty vegetables or hands, it is also important to wear gloves, wash hands and vegetables effectively, and remove outer leaves and peel vegetables to remove contaminant risk.[109] Because the health risks from eating the produce itself are generally quite low, unless crops are grown on heavily contaminated soil,[110] a combination of good soil management, including regular use of compost and manures, and attention to hygiene, should keep exposure risks low.

Soil contamination is not just a problem for urban gardeners. Many towns and cities contain vast areas of derelict wasteland, where industrial activity once flourished. This land is often referred to as brownfield, defined as previously developed industrial or commercial land that has potential to be redeveloped. Areas of brownfield can be considerable. A report on the scale of European brownfield estimated that there is as much as 65,000 hectares in England's towns and cities, which is roughly equivalent to 90,000 football pitches, and 128,000 hectares in Germany, 20,000 hectares in France, and a staggering 800,000 and 900,000 hectares in Poland and Romania respectively.[111] The exact area of brownfield in the USA is not known, and varies considerably across its cities and towns, but a recent national report indicated that there could be as many as 400,000 sites.[112]

Cleaning up and recycling brownfield land into productive use, such as for housing or commercial developments, is a major priority for many nations and is seen as an important step for stalling urban sprawl into surrounding green belts and country-side. Such developments raise many political challenges, which I don't want to go into here. But they also bring challenges that relate to the soil, which is often tainted by chemicals and rubble, and lacking in organic matter and the nutrients that are needed for plants to grow. These challenges can be considerable. The remedi-ation of land for the 2012 London Olympics, which covered 200 hectares of derelict industrial land, is an excellent example of what can be achieved. This was the UK's largest ever soil-cleaning operation, and involved excavating and cleaning up for reuse over 2 million cubic metres of soil, which was contaminated with heavy metals, solvents, various organic toxins, and rubble. To do this, the company charged with the task of remediating the land set up two on-site 'soil hospitals' and several soil-washing plants to test, treat, and recycle contaminated soil.[113] The development was hailed a great success, and what was once a derelict industrial wasteland is now a vibrant and sustainably landscaped area of London supported by clean and healthy soil. This is just one example, but across the world many major soil remediation pro-jects have provided an essential foundation for the successful redevelopment of once derelict and polluted land.

NIGHT SOIL

Before the provision of toilets and public sewers in the late nineteenth century, a common practice in English cites, and many other cities of the world, was the collection of human refuse, which was transported to land surrounding the city and used as a fertilizer (Figure 13). This was especially common in large cities such as London and Manchester, where rapid population growth during the nineteenth century generated vast amounts of human waste. To deal with this waste, which was politely called night soil, 'night soil men' or 'scavengers' would collect excreta during the night from backyards and houses, and transport it in carts and boats to the countryside to fertilize the land. In Manchester, historic records document night soil being collected in the late nineteenth century from the city and loaded into barges at Salford Docks, from where it was shipped to the outskirts of the city. It was then uploaded onto tramways and taken by train to areas of boggy land,

FIGURE 13 Night soil collectors emptying their carts onto land

such as Chat and Carrington Moss, which have since been consumed by the city. Years of this practice caused these areas to change from being wet, unproductive land to fertile agricultural land, which they still are today. Records of this practice also come from other cities such as Liverpool, where during the nineteenth century 'night soil men' would have the unenviable job of carting night soil to the canals, where it was transported to the surrounding Lancashire countryside to be used as a fertilizer on farmland.

The use of night soil can reap major benefits for the fertility of farmland soil, but it also creates problems. Not only is its use considered by many to be a degrading and unpleasant activity, but also the use of untreated human wastes to fertilize land can cause major human health problems because of the many pathogens that excrement contains. Because of this, the use of night soil has been banned in many countries. But in some, such as China, its use as a fertilizer, and source of biogas, is still practised today, although the millions of tonnes of night soil being produced in Chinese cities each year has to be first treated to control for pathogens before it can be used on land. In fact, the use of night soil in China, and other countries such as Korea and Japan, dates back centuries, and is arguably one of the reasons why their soils have remained healthy and productive for so many years. Indeed, at the turn of the twentieth century, the American soil scientist F. H. King wrote of the astonishing amounts of human waste that were carefully collected from cities across China, Korea, and Japan, which returned significant amounts of nitrogen, potassium and phosphorus to the soil.[114]

I began this chapter writing about how, for most, soil doesn't feature much in urban life. City dwellers might come into contact with soil in their garden, or when potting plants, or while playing a sport. But I suspect that most city dwellers, if asked, wouldn't consider soil as playing a key part in urban life. This is clearly not the case, however, and in writing this chapter, I have also become more aware of the many roles that soil plays in urban life: it filters the water after storms, helping excess water drain away; it stores carbon and nutrients and helps to regulate the heat and quality of urban air; it binds and breaks down the myriad pollutants that the industry of cities yields; and it provides a foundation on which city dwellers grow their crops and play their sports. In fact, soils play a major role in cities and while much urban land is covered in asphalt and bricks, a great deal is not, and much of this land is underlain by healthy soil. Next time you walk through a city, pause to consider not only the bricks, asphalt, and mortar which cover much of the land, but also the many green spaces and human activities that depend on healthy urban soil.

5

Soil and War

Bent double, like old beggars under sacks,
Knock-kneed, coughing like hags, we cursed through sludge
                                                    Wilfred Owen

My first visit to a battleground was during a family holiday to Scotland. We were staying in Applecross, a small, isolated village on the west coast of the Scottish Highlands that looks over the sea towards the Island of Raasay. On the way back we passed through Inverness, the most northerly city in Scotland. To break the long journey we decided to stop off at Culloden Moor, the site of the Battle of Culloden in 1746, between the Government forces, which were mainly English, and the Jacobite army, made up of Scottish Highlanders led by Bonnie Prince Charlie. I had never visited the site before, but I recall thinking that it was an odd place for a battle; it is exposed moorland and the ground is rough and boggy, which would be difficult ground on which to go to war. I later learned that Bonnie Prince Charlie's choice of this site for battle was catastrophic; not only did the exposed ground leave the Jacobite forces vulnerable to the superior artillery of the Government forces, but also the boggy soil hampered their attack, rendering them even more exposed. These factors led to the slaughter of the Jacobite forces and the collapse of the Jacobite campaign. I don't know exactly how much the boggy soil contributed to the outcome of this war but it certainly played a part.

For centuries, soil has played an enormous, and often unexpected, role in the outcome of war. War can also leave lasting and often irreversible scars on soil, leaving it churned, riddled with battle debris and bodies, polluted with heavy metals, toxic dioxins, oil and radioactivity. In many cases, it is left unusable. War can also indirectly affect the soil, for example through the need in Britain, during the Second World War, to cultivate gardens and city parkland for food. And the current growing demand for food, coupled with environmental pressures related to climate change, will place increasing pressure on soil, potentially leading to future wars. This chapter will look at how war is affected by and how it affects soil.

MUD AND THE FIRST WORLD WAR

Much has been written about the close connection between warfare and the soil landscape. But perhaps the most dramatic and emotive example of soil playing a major role in conflict concerns the First World War. The scene of the First World War was the Western Front, which stretched over 700 kilometres from the Belgian coast in the north, southwards to Pfetterhouse on the border of Switzerland. Many battles were fought along this front, but the most costly were the Battles of the Somme and Passchendaele, which respectively left 1 million and 600,000 dead, and the Battle of Verdun with an estimated 700,000 casualties.

FIGURE 14 During the First World War, intense artillery bombardment, combined with heavy rain and poorly drained soils, created a quagmire, causing havoc for military operations and making the lives of soldiers hell.

The most obvious impact of soil on these battles was through the swathes of knee-deep mud that surrounded the front line (Figure 14). This mud emerged from intense artillery bombardment, combined with heavy rain and often poorly drained soils, and it caused havoc for military operations and made the lives of soldiers hell. I recently came across an article in my local paper, the *Manchester Evening News*, which painted a grim picture of the overwhelming influence of mud in this war. It included a quote by Major Chesnutt-Chesny, who led the 2/6th Lancashire Fusiliers into battle on Passchendaele Ridge on the morning of 9 October 1917. He wrote:

Men stumbled forward through mud and slime. At every step they sank over the boot-tops in sodden soil; not a few sank waist-deep in the yawning, water-filled shell holes—but they staggered on with the grim persistency...[ra1][115]

As we will see, many soldiers who fought in the First World War wrote of the mud, and it left an overpowering impression in the minds of many of those who survived.

The Battle of Verdun, which was fought between the Germans and French in 1916, occurred under terrible conditions. Not only is the area one of the wettest in France, with long cold winters and persistent heavy rain, but also the soil of the area is heavy clay, which becomes bogged down and waterlogged when wet in the winter. As a result of these conditions, the heavy artillery bombardments between the Germans and French, which were a key feature of this war, created a quagmire. This hampered attacks and led to many soldiers drowning in the very mud and craters that their own sides made. Joseph Hupey, in a paper on the long-term impact of this battle on soil development,[116] described how the exchange of artillery, which lasted for almost a year, pulverized the landscape into what many soldiers described as 'something from another world' or what pilots flying above would refer to as 'the surface of the moon'. The scars of this war on the landscape and its soil are still clear to see today.

The situation was even worse at the Battle of Passchendaele, a major offensive by the British and their allies against the Germans. This battle is known as the Third Battle of Ypres, and began in July 1917 and ended in November of the same year. It became infamous not just for the enormous and wasteful loss of life, but also for the mud. The objective of the battle was to take control of the strategically important ridges that the Germans held to the south and east of Ypres, a small town in West Flanders, with the ultimate aim of forcing them to withdraw. Much has been written about the strategic background to this battle, the battle itself, and its outcome. Here, I will focus on the mud and how it played such a major role.

The battle took place on lowlands that had been reclaimed from marsh for agriculture by a myriad drains, small waterways, and canals; the Germans were positioned on the higher ridges that formed a natural amphitheatre from which they could observe all that went on below. Although the soils of the battleground were sandy in texture, the area is underlain by heavy clay. This, combined with the high water table, made the ground sodden, especially during periods of heavy rain.

The battle began with an intense bombardment, designed to weaken and demoralize the Germans. The words of a British soldier provide a graphic description of the scene:[117]

> It was a terrible sight, really awe inspiring, to see the barrage playing on the German front lines before we went over. It was an inferno.

But this bombardment was so intense it destroyed the drainage systems and churned the land into a crater-filled quagmire. Then, as if things couldn't get worse, when the Allied infantry launched their attack the rain began to fall, far more heavily than for some thirty years, and as the rain fell, the soil turned into liquid mud.

The mud caused havoc. It clogged rifles, bogged down field guns, flooded trenches, swallowed up tanks, and slowed down rescue operations as stretcher-bearers struggled through waist-deep mud. It took four, or even six, men to haul a stretcher to safety, when it would normally take two, and a journey of as little as 200 yards could take two hours of struggle through the lashing rain and sucking mud.[118] Soldiers and horses became trapped in the mud, and many drowned in it, including the wounded. Consider an extract from the memoir of Lieutenant Annan, of the 1st/9th Royal Scots Regiment:[119]

> It was the wounded, lying there sinking, and this liquid mud burying them alive, running over their faces into their mouth and nose... We couldn't understand why, in the name of God, anyone had ordered an attack like that over terrain like that. It was impossible.

As well as severely hampering military activity, the mud also created anxiety and disillusionment among soldiers. These emotions emerge from the memoirs of soldiers. Santandu Das, in his book *Touch and Intimacy in First World War Literature*,[120] refers to the words of gunner Jack Dillon:

> The mud there wasn't liquid, it wasn't porridge, it was a curious kind of sucking mud. When you got off this track with your load, it 'drew' at you, not like quicksand, but a real monster that sucked at you.

He also refers to an article from the front-line newspaper *Le Bochophage* which describes how

> the mud watches, like an enormous octopus. The victim arrives. It throws its poisonous slobber out at him, blinds him, closes round him, buries him... For men die of mud, as they die of bullets, but more horribly.

In the same book, Das also highlights how mud brought home the ugliness of war. He refers to a solider that wrote:

> What a life, Mud, earth, rain. We are saturated, dyed, kneaded. One finds dirt everywhere, in pockets, in handkerchiefs, in clothes, in food. It is a haunting memory, a nightmare of earth and mud.

The horror and anxiety that mud caused was such that soldiers referred to mud as the worst hell, even worse than the shelling. As written in the *Le Bochophage* at the time:

Hell is not fire, that would not be the ultimate in suffering. Hell is mud.

The Battle of Passchendaele ended in November 1917 with the British and Allied forces taking Passchendaele village from the Germans. But the cost, in terms of casualties on both sides, was enormous, and many bodies were never found; they lie where they sank into the mud, along with other debris of war.

THE MUD MARCH AND WATERLOO

The First World War wasn't the only war in which mud played a major role. Only recently, in a major offensive in Afghanistan, British Royal Marines trudged through miles of mud while fighting insurgents at close quarters, to capture Taliban strongholds near the town of Nad-e-Ali. Reporting on the offensive, the BBC commented: 'They fought knee-deep in mud during First World War-style trench battles.' Two battles that stand out in history as being linked to mud, however, are the Mud March of the American Civil War, and the Battle of Waterloo, where the Duke of Wellington defeated Napoleon.

Prior to the Mud March, the Union army, under the command of General Ambrose Burnside, suffered a devastating defeat to the Confederates, led by General Robert Lee, at the Battle of Fredericksburg, on 13 December 1862. In an attempt to restore his reputation after the defeat, and raise the morale of his troops, Burnside decided to launch another offensive on Lee's army. The plan was to cross the Rappahannock River upstream of the

Confederates and attack them from behind. The offensive began on the morning of 20 January 1863, in mild weather; but that night the rains started to fall heavily, and by the morning of the 21st, the soil was waterlogged and had turned into a quagmire. Lieutenant Theodore Dodge described the conditions in his *Journal:*[121]

> You have no idea of how soon the roads turn from good to bad here in Virginia ... A clayey soil is hard and the very best for marching on in favorable weather, but let it rain but an hour and troops and wagons march over the road, and the mud is worse than anyone who has not been in Virginia can conceive of.

The rains continued to fall and as the mud got worse, horses, wagons, and heavy artillery sank, and soldiers struggled to move forward. The mud was so bad that what should have taken an hour or so took days. Men were exhausted. A reporter for the *New York Times* commented:

> Horses and mules dropped down dead, exhausted with the effort to move their loads through the hideous medium. One hundred and fifty dead animals, many of them buried in the liquid muck, were counted in the course of a morning's ride.[122]

The advance was aborted on 24 January and, on the next day, Major General Joseph Hooker was brought in to replace the humiliated Burnside.

A similar situation emerged at the Battle of Waterloo, 1815, which was fought between the French, under the command of Napoleon Bonaparte, and the Allied armies, led by the Duke of Wellington from Britain and General Blücher from Prussia. Several factors contributed to Napoleon's defeat, but the battleground soil played its role. Wellington's army was positioned on higher, drier ground, whereas Napoleon's army was on lower, wetter ground, with sodden soil. To make things worse, it rained heavily the night before the battle and the lower ground became even more sodden and mud-ridden, making it even more difficult for artillery and cavalry to move. Napoleon decided to delay the start of the battle to allow the ground to dry a bit. But this delay meant that Napoleon lost his advantage, and the deteriorating ground resulted in the French having to take the battle to Wellington uphill through mud. Other factors obviously played a role in the outcome of this battle and Napoleon's ultimate defeat; but, as in so many other battles, so did the mud.

TUNNELS OF WAR

While researching for this book, I came across a remarkable piece of information: during 1917 and 1918, more people lived below-ground in the Ypres area of the Western Front, where the Battle of Passchendaele was fought, than reside aboveground in the town today. I looked into this more, and discovered that during the war, around 50,000 tunnellers, sewer workers, and miners were engaged in mine warfare below the battlefields of this area; a secret struggle beneath no man's land.[123]

Tunnelling has been part of warfare for centuries. It first emerged as a key part of siege warfare, as a way of effecting surprise. As far back as 2000 BC, the Egyptian Army used tunnels to bring down fortress walls by undermining, and later in 435 BC, at the siege of Fidenae, the Romans used tunnels to penetrate directly into the heart of the fortress. More recently, tunnel warfare was used to great effect during the First and Second World Wars, and the Vietnam War, where the Vietcong used a vast complex of tunnels to hide from the American forces and to attack them without warning. I don't wish to go into the history of tunnel warfare here; rather, I want to illustrate, through selected examples, how knowledge of soil and underlying geology are central to the art of tunnel warfare. Successful military mining, as noted by the historian Peter Barton and his colleagues in *Beneath Flanders Fields*,[124] demanded precisely the same preconditions as it had for millennia: that the position of the objective was fixed, that it was within reach, that it would remain immobile, that there was enough time to carry out the work, and that the ground between the attacker and the objective was mineable. In other words, that the soil and underlying geology could be easily worked and excavated.

Tunnel warfare, and the required understanding of soil conditions and geology, played a major role in the First World War (Figure 15). The main aim of tunnel warfare was to place mines

FIGURE 15 Knowledge of soil and underlying geology were central to the art of tunnel warfare during the First World War.

under enemy positions, creating huge explosions that killed many enemy soldiers and undermined their defensive positions. A notable example of the scale of such explosions, and their devastating impact, was at the Battle of Messines, June 1917. For a year before the battle, thousands of British, Australian, and Canadian miners dug a series of tunnels beneath German lines on Messines Ridge, where they then placed 455 tonnes of explosives, which created huge explosions that devastated the enemy lines and killed around 10,000 German solidiers. Peter Barton and his colleagues referred to the words of an anonymous tunneller who witnessed the scene:

> Suddenly, all hell broke loose. It was indescribable. In the pale light, it appeared as if the whole enemy line had begun to dance, then, one after the other, huge tongues of flame shot hundreds of feet into the air, followed by dense columns of smoke, which flattened out at the top like gigantic mushrooms.

The Ypres salient was ideally suited to tunnelling because of the underlying clay: the Ypres Clay, or *Argile de Flandres*. This clay is blue in colour, turning brown when exposed to air, and is very plastic, making it an ideal medium for driving tunnels. It is actually of similar nature to London Clay, which is also ideal for tunnelling. This is one of the reasons why the London Tube railway network expanded so rapidly. But a key characteristic of heavy clay is that it absorbs water when wet, which makes it expand, and when it dries it shrinks. This shrink-swell activity created major problems for tunnellers, because the expanding clay placed considerable pressure on timber supports. This was especially a problem when tunnelling at depth, because the introduction of moisture to dry Ypres Clay, via the damp air of a newly driven tunnel, caused rapid swelling of walls, floor, and roof. The clay would eventually settle and cease swelling, but sometimes it was so severe that metal girders were needed, and those with timbers needed constant maintenance.[125]

Another problem was that the underlying geology around Ypres wasn't uniform, being overlain in places by complex sediments of sand and silt. This created an often unpredictable soil to tunnel, and such variations in soil could determine the success or failure of a mine. Passchendaele Ridge on the German front line presented such a complexity of soil types. The underlying base is Ypres Clay, which is overlain by sandy clay, the Paniselian Clay, which was known as 'bastard blue clay' to the miners, and then an upper layer of saturated sand, which was capped, in places, by moist sandy clay. This presented a complex mixture of substrates to mine, each with its own challenge. The Ypres Clay was easy to tunnel, but prone to expansion, whereas the Paniselian Clay was a tunnellers' dream, because the mix of sand and clay reduced the plasticity of the soil, and lowered its capacity to expand. In contrast, the sand, which was found on the ridges, was extremely difficult, and often impossible, to mine, being waterlogged and like quicksand in nature. These saturated sands were more of a problem for the Germans, being positioned on the higher ground where they occurred. This presented an ironic reversal of fortune for the Germans: when mining began, the tactical advantage of holding higher ground, in terms of better observation and artillery fire, became a major liability.[126]

The Western Front straddled two main geological zones: the wet clay of Flanders and the free-draining chalk of the south. The chalk is soft and porous, and hence easy to dig. And as a result, underground warfare was central to the war effort in this region, most notably at the Battle of Vimy Ridge, 1917, a major offensive between Canadian forces and the Germans. The strategic aim of the battle was for the Canadians to take control of the high ground of Vimy Ridge that the Germans occupied. The advantage of holding Vimy Ridge for the Germans was obvious: it is an escarpment that lies some 100 metres above its surroundings, providing sweeping views of northern France and the Douai Plains.

Tunnelling was used by both sides for the placement of explosives beneath enemy lines, and to protect advancing soldiers from gunfire, and the geology and soil conditions at Vimy played a major role in this. Not only were the soils easy to dig, but also the chalk was very stable and dry, so tunnels were more self-supporting than at Flanders. The underlying geology, however, also brought problems. In particular, the fracture orientations of the chalk meant that sections of the roof were prone to collapse, or slabbing, during excavation, so roof width needed to be kept to a minimum. Also, because the water table fluctuated by some 10 metres between winter and summer, there was much potential for tunnel flooding and for surface soils to become waterlogged and, being heavy clay, a morass of mud. Other hazards of the chalk were that it is hard, so the action of digging could not be done in silence, as could be done in the clays of Flanders. Because of this, the element of secrecy was harder to control. And, as noted by Barton and colleagues, the spoil from underground workings gleamed brilliant white in sunshine, so if left unhidden, they were an obvious clue to aerial observers that mining was taking place.[127]

Tunnels also played a major role in the Vietnam War, where they were used by the Vietcong to move guerillas from place to place, and to shelter from American bombs and artillery. The maze of tunnels was also used to launch surprise attacks and to spy on, and seize weapons from, the Americans. The most extensive tunnel network was in the Cu Chi district of South Vietnam, which was a stronghold for the Vietcong throughout the war.[128] There were hundreds of kilometres of tunnels, which stretched from Saigon—now called Ho Chi Minh City—to the Cambodian border. The tunnel network harboured an underground world, with living areas, hospitals, storage depots, military headquarters, and it connected villages, districts, and even provinces. As noted by General Westmorland, who commanded the American forces in Vietnam from 1964 to 1968: 'the Viet Cong; they were human moles'.

The network of tunnels was first constructed in the 1940s during the Indochina War when the French reoccupied Vietnam. It then expanded in the 1960s during the Vietnam War when the Vietcong were forced underground because of the American's extreme use of aircraft, bombs, artillery, and chemicals in this district.[129] Indeed, Cu Chi became the most bombed, shelled, gassed, and defoliated area in the history of warfare. And, because the Vietcong went underground, the war dragged on to the point that the Americans decided it was unwinnable.

The soil conditions in Cu Chi were highly conducive to tunnelling, providing the Vietcong with a huge environmental advantage. The soils were mainly lateritic clays, which are highly weathered, rusty red soils that occur across the humid tropics. A key feature of these soils is that they are very hard, porous, and, unlike the heavy clay soils of Flanders, stable when wet; as a result, they provide a very stable material for tunnelling. In fact, the Scottish physician Francis Buchanan-Hamilton first used the term 'laterite' in 1807, when working in India, to describe a material that was soft enough to cut *in situ*, but hardened when exposed.[130] The material was being quarried for bricks; hence the term 'laterite', which he derived from the Latin *later*, a brick. The roots of a variety of trees further stabilized the tunnels, and the water table was at 10 to 20 metres' depth, which again made for ideal tunnelling conditions; as Captain Linh, the commander of the Vietcong guerrillas of the Cu Chi district, put it: 'We could not have expected better conditions.'[131]

Although ideal for tunnelling, the disposal of soil created problems for the Vietcong. Not only did they have to dispose of thousands of tonnes of soil from the tunnels, but they also had to do this in a way that wasn't obvious to the Americans. Solutions included pouring excavated soil into streams and using it to make furrows for potatoes, basements for houses, and banks for combat trenches. And when the American B-52 bombing raids first began, soil was shovelled into new craters. The Vietcong also came up with ingenious ways of secretly transporting soil away from tunnels, such as in a crock, which was a small pot, the size of a coffee pot, used by women to carry fish sauce. The key point of all of this was that excavated soil should never be left heaped in mounds, which would reveal the presence of tunnels to the Americans.

Another aspect of war in which tunnels and the soil have played a role is in the escape from prisoner-of-war camps. In many cases, tunnelling was the only option for escape, aside from daring attempts to hide in transport vehicles or cut through the barbed wire fences that surrounded prisoner-of-war camps. One of the most famous attempts at escape was that depicted in the movie *The Great Escape*, which was based on a book of the same name by ex-prisoner Paul Brickhill, in which officers of the Allied air forces attempted to escape from a Nazi prisoner-of-war camp, Stalag Luft III, during the Second World War. Soil played a central role in the story of *The Great Escape*, mainly due to the problems encountered in disposing of soil from the tunnels. The problem was the colour of the soil, in that its surface was grey, whereas the deeper soil, which was excavated from the tunnels, was bright yellow sand. Not only was this deeper sand unstable, making tunnels prone to collapse, but the contrast in colour made it difficult to conceal the subsurface sands, which, as one ex-prisoner put it, 'shine like drops of gold' if spilled on the camp floor.[132] This presented a major problem for the prisoners, who had to come up with ingenious ways of disposing of it, such as dispersing the sand into the grey surface soil of gardens and from bags suspended within trouser legs.

The soil at Stalag Luft III was a podzol, formed on freely drained sand and gravel. Podzols are among the most striking of soils, because their horizons differ markedly in colour: below the surface layer of dead litter and humus, there is typically a bleached grey layer, beneath which there is a deeper layer of vibrant orange-red soil. These layers are formed as a result of the stripping of minerals, especially iron and aluminium, from the upper soil, and their deposition deeper in the soil. This removal of iron from the upper soil occurs most rapidly when water can run through soil easily, as is the case in the sandy soils of Stalag Luft III. Over time, this leads to the formation of a layer that is bleached in appearance, called the 'ash' horizon, whereas the deposition of iron deeper in the soil forms a subsurface layer that is deep orange-red, or rusty, in colour. It is these processes of podzol formation that caused the problem for the prisoners, but were to the advantage of the Nazi guards of Stalag Luft III.

BOMBTURBATION

Warfare and all its horror can leave many scars on soil, both visible and invisible. The most visible signs of war result from explosive munitions, including aerial bombs, propelled explosives, and *in situ* mines, which crater and mix the soil, often to considerable depth. Soil scientists have studied in great detail how explosive munitions impact soil development, and two American scientists, Joseph Hupey and Randall Schaetzl, even introduced the term *bombturbation* to describe the cratering of the soil surface and mixing of the soil by explosives.[133] They argued for the need for this term, given the unique way explosives disturb the soil and the incomprehensible intensity of this disturbance that has occurred across the world. Literally millions of craters were produced during the First World War, and during the Second World War almost 1.5 million tonnes of bombs were dropped on Europe, and 0.5 million tonnes of bombs were dropped on Germany alone by American bombers. As Hupey and Schaetzl point out,[134] these numbers pale in comparison with the 14 million tonnes of bombs dropped over Indochina by the Americans during the Vietnam War.

The most detailed account of bombturbation comes from the First World War battleground of Verdun, France, which covered almost 30,000 square kilometres, and is one of the most heavily shelled of all time. Hupey and Schaetzl visited the battleground in 2004, eighty-eight years after the battle. They dug a number of pits to examine soils in undisturbed areas next to craters, as well as in disturbed areas within craters.[135] They discovered that since the time of the battle, when no recognizable soil would have remained, soils within craters had developed thick surface layers of organic matter, and the bedrock that was exposed during the war had weathered to form shallow mineral soils. They also discovered that craters have acted as focal points for water run-off, and because of this, plant debris and sediments have washed into craters, which has accelerated the recovery of soil. Another intriguing observation was a high level of worm activity in craters, which has enhanced the breakdown of organic matter and mineral weathering, thus speeding up the restoration of soil. The scars of warfare still remain at Verdun, and will for centuries to come. But this study showed that soils have a remarkable ability to recover from catastrophic disturbance caused by war.

Explosive munitions have also left their mark on soils of Vietnam, especially at Khe Sanh, the site of a major offensive between the North Vietnamese Army and the US Marines. Here, in 1968, the US Air Force launched a massive aerial bombardment campaign, called Operation Niagara, which scarred the landscape to the extent that soldiers described it as a 'moonscape' or 'another world'. And, as at Verdun, the scars remain today: hilltops are littered with craters, and while vegetation has recovered in places, the patterns of recovery have been affected by the soil conditions within craters. In general, soil within craters tends to be wetter than on the rims, because of the collection of water in crater bottoms, and woody vegetation has regenerated inside craters, but not outside where it is too dry.

AGENT ORANGE

Warfare leaves its mark on the soil in other ways. The use of chemicals, for example, can leave soil indefinitely polluted and, as in Vietnam, the legacy of extensive use of toxic herbicides remains

FIGURE 16 US forces used enormous quantities of toxic herbicides during the Vietnam War to defoliate forests and destroy crops. These herbicides contained dioxin, which has left a legacy of contaminated soil.

today (Figure 16). US forces used enormous quantities of herbicides during the Vietnam War to defoliate forests and mangroves, to clear perimeters of military installations, and to destroy crops in order to decrease enemy food supplies. Around 2.6 million hectares of land was sprayed with herbicides during the Vietnam War, and during its peak between 1967 and 1969, more than 16 million litres were sprayed annually.[136] Several herbicides were used, but the one that gets most attention is Agent Orange. The main problem with this herbicide was that it contained 2,4,5-trichlorophenoxyacetic acid (2,4,5-T), which was contaminated with the toxic chemical 2,3,7,8 tetrachlorodibenzo-p-dioxin (TCDD), or dioxin.

The effects of direct human exposure to Agent Orange, and the accompanying dioxin, were horrifying. Even today, millions of

Vietnamese and US veterans are plagued by illness, and there are claims that thousands of children continue to be born with horrific facial deformities due to its use. A key reason for these continued effects is that the dioxins persist in soil for many years, and because they have a strong affinity for fat, they also accumulate through the food chain to fish and livestock, which are consumed by humans. A study in 1995 of 3,200 Vietnamese people found much higher levels of TCDD in the blood, breast milk, and adipose tissue of people living in areas sprayed with Agent Orange compared to those living in unsprayed areas of Vietnam.[137] Also, after the war had ended, US war veterans were found to have high levels of TCDD in their blood and adipose tissue due to exposure during the war.[138]

Contamination is worst around the former US military bases where the herbicides were stored. Here, soil concentrations of TCDD remain extremely high, most likely due to spills of Agent Orange that were common during the war. A study carried out between 1996 and 1999 in the Aluoi Valley of central Vietnam, which had three US Special Forces bases and was extensively sprayed, illustrates this point.[139] They found that soil TCDD contamination, and the subsequent transfer of TCDD through the food chain to humans, was greatest in areas of former military installations, where concentrations were especially high from herbicide spills. Similarly, soils taken from a nearby former military installation at Bien Hoa, where there was a reported spill of 7,500 gallons of Agent Orange in 1970, were highly contaminated with TCDD, as was the blood of people in the area.[140] Here, the levels of TCDD in soil were several orders of magnitude higher than what would be considered acceptable anywhere else. Another source of soil contamination was emergency dumps during aborted spraying missions, where the herbicide was jettisoned in about thirty seconds, rather than the usual four to five minutes, and several herbicide-loaded aircraft crashed.

Why does TCDD remain in soil for so long? The main reason is that it binds tightly to soil organic matter, making it less immobile, and also few soil microorganisms are equipped to degrade dioxins such as TCDD, so its breakdown is extremely slow. Because of this, clean-up of TCDD-contaminated soil is not simple, and involves either sealing off contaminated sites with a concrete cap, or removing soil from a site for treatment to destroy the dioxin. This latter approach involves digging up vast quantities of contaminated soil, which is then placed in storage tanks and heated to extreme temperatures to destroy the dioxin. This is extremely costly and time-consuming, but it is the only way to rid soils of dioxins.

DEPLETED URANIUM

A legacy of more recent war is soil contamination with depleted uranium,[141] which has been used as an armour penetrator due to its extremely high density and hardness. Depleted uranium penetrators can slice through the armour plating of modern tanks with devastating effect, and they were used extensively during the 1991 Gulf War, the 1999 Kosovo conflict, and the 2003 Iraq War. Typically, aircraft equipped with one gun could fire upwards of 3,900 rounds per minute, and a burst of fire of just two to three seconds would involve 120–95 rounds. Relatively few of the penetrators actually hit their target, and many of those that missed would penetrate the soil, often to several metres' depth, where they would remain embedded. During the Gulf War, for example, the American and British forces fired more than 300 tonnes of depleted uranium, and given that many penetrators missed their targets, much of this depleted uranium is now buried in soil.

Depleted uranium causes harm to human organs not only because of its toxicity as a heavy metal, but also due to the release of radiation. Because of this, there is much concern about the legacy effects of depleted uranium for human health in war zones, and although refuted, it has been claimed that it is the cause of leukaemia and other health effects among troops and civilians involved in the conflicts.[142] The main health risk is through direct inhalation of depleted uranium dust created when a penetrator hits its target; being very fine, this dust is easily carried by wind. But another worry is that the penetrators that enter the soil will corrode and contaminate the soil.

Studies conducted in Kosovo and Kuwait to test for post-war soil contamination by uranium[143] have found that, while soil uranium concentrations are generally normal, soils hit by rounds contained literally hundreds of thousands of microscopic particles of depleted uranium, which created hotspots of soil contamination. I later discovered that colleagues at the University of Manchester have been exploring how such small fragments of depleted uranium from penetrators corrode in soil, and how they influence the resident microbes. The team performed a series of controlled experiments and discovered, to their surprise, that soil microbes were ineffective at breaking down depleted uranium.[144] Of even more concern, they also found that the corrosion of depleted uranium fragments in soil actually had a negative impact on microbial diversity in soil, with certain key groups of bacteria being completely wiped out.[145] They didn't test whether this impaired key soil functions. But given that soil microbes play such an important role in building soil fertility, such declines in microbial diversity around the corroding depleted-uranium fragments could have a negative impact on soil fertility in their immediate vicinity.

FALLOUT FROM NUCLEAR WEAPONS

Soil contamination with depleted uranium is very localized and confined to the zones of conflict where it was used. In contrast, the testing of nuclear weapons, and their use in war, has completely changed the global radiation environment. In fact, there is now no place on Earth where the signature of nuclear bomb testing cannot be found in soil. Nuclear bomb testing first started in New Mexico in July 1945, and in August of the same year, American B-52 aircraft dropped nuclear bombs on the Japanese cities of Hiroshima and Nagasaki with apocalyptic effect. After the Second World War, various nations, including the Americans, Russians, Chinese, French, and British, continued extensive testing of nuclear weapons at numerous sites across the world, although they were never used again in war; over a period of around forty years, from the first bomb test in 1945 to the 1996 Comprehensive Nuclear-Test-Ban Treaty,[146] which saw an end to bomb testing by most nations, upwards of 2,000 nuclear tests had been performed across the world.

Most will be familiar with the vast mushroom that forms when a nuclear bomb explodes. Such mushrooms produce vast amounts of radioactive fallout, including plutonium and deadly fission products such as caesium (^{137}Cs), and other debris, which can be spread and deposited over literally thousands of miles, depending on prevailing winds, exposing humans to radiation not only through direct inhalation, but also through consuming contaminated food and water. There is no shortage of graphic descriptions of fallout and its horrific effect from Hiroshima and Nagasaki, where the explosions caused ferocious firestorms and 'black rain' pouring down over the city and surrounding areas. This 'black rain' was a mixture of carbon residues from the fires and radioactive material, and it fell to the ground as a sticky, dark, and dangerously radioactive material causing unimaginable harm.

A number of studies have measured the soil legacy of radiation in and around Hiroshima and Nagasaki as a result of the 1945 bombs. In one study, a team of scientists from Canada, Japan, and France collected soil samples from areas around Nagasaki forty years after the explosion.[147] Some samples were taken from surface soil just to the east of the hypocentre where the 'black rain' fell, and the rest were collected from up to 100 kilometres to the east to test how the fallout had spread. They found that the peak deposition of plutonium (^{240}Pu) was not at the hypocentre, but some 2.8 kilometres to the east, where the 'black rain' poured down on the city. Moreover, the concentration of plutonium (^{239}Pu and ^{240}Pu) at this point was about 100 times greater than the background level expected for this part of the world. Further to the east, levels of plutonium in surface soil dropped precipitously at 5 kilometres from the hypocentre, and at 100 kilometres they were minimal. These results showed that even after forty years, soils close to the hypocentre, where the 'black rain' fell, were still contaminated with plutonium. Subsequent studies of forest soils around Nagasaki showed that even after sixty years, soil contamination with plutonium remained from the detonation of the 1945 nuclear bomb.

A major problem with radionuclides is that many decay very slowly.[148] Because of this, they hang around in soil for many years. Another problem is that they can transfer from soil to plants, and then to animals that eat them. Although not related to conflict, our knowledge of how radionuclides, especially caesium, behave in soil and pass through the food chain took a major leap forward following the Chernobyl nuclear reactor accident in April 1986. Chernobyl was the worst nuclear power plant accident in history, and created a vast plume of radioactive fallout that drifted across large parts of the Soviet Union and Europe, causing caesium contamination as it fell. The highest radionuclide deposition occurred in Belarus, the Russian Federation, and Ukraine, but high fallout also occurred in other parts of Europe.

At the time of the accident, I was at home, on the edge of the Lake District, one of the areas in England worst hit by Chernobyl fallout. This was because of a heavy downpour that happened in the region as the radioactive plume passed by. The initial view of the author-ities was that the caesium would become strongly bound to organic matter in the peaty soils that cover much of the region, especially on higher ground, rendering it unavailable to plants and grazing animals. Yet, within weeks of the accident, levels of caesium in sheep rocketed from grazing on contaminated plants. As a result, a ban was immediately placed on the movement and slaughter of sheep, initially for three weeks, but then indefinitely. Ten years after the accident, restrictions remained on several Cumbrian sheep farms, and on others in the UK, and in 2012, twenty-six years after the accident, the last restrictions were eventually lifted.

One of the reasons that the radioactive contamination stays around for so long is down to the soil. When caesium is deposited on the ground, much of it is washed into the soil, where it either becomes tightly bound to organic matter and clay minerals, or remains dissolved in soil water. The caesium in soil water can be taken up immediately by plant roots and transferred to shoots, causing contamination of plants and the animals that feed on them. Caesium that is bound to organic matter, however, is released slowly and taken up from their roots, rather like a drip-feed of radioactivity. One of the surprising findings to emerge from research after Chernobyl was that soil fungi can also act as a major store of caesium, holding as much as 20 per cent or more of the total amount found in soil. Mycorrhizal fungi, which attach to plant roots, are especially effective at accumulating caesium in their tissues, and also act as a pipeline for the transport of caesium to plant roots, thereby promoting contamination of plants and animals that feed on them.[149]

Following Chernobyl, it was also found that soil fungi transport large amounts of caesium to their fruiting bodies, or toadstools, which in some parts of the world are commonly eaten by humans. As a result, following Chernobyl, concentrations of caesium in toadstools in many parts of Russia and Europe have been found to be higher than in many other foodstuffs, and in some regions the eating of mushrooms is a major source of caesium intake by humans. In rural regions of Russia, for example, the consumption of wild mushrooms was a major source of contamination of humans following the disaster, and was the reason behind a surge in caesium activity in humans in autumn, when fruiting bodies appear.[150]

DIG FOR VICTORY

During war, soil also plays a major role in civilian life, especially through the need for home-grown produce when food supplies are cut off. An example of this was the British 'Dig for Victory' campaign during the Second World War, which encouraged civilians to convert their lawns, flower beds, sports pitches, and

FIGURE 17 During the Second World War, the 'Dig for Victory' campaign encouraged British civilians to take their spades to the soil and convert their lawns, flower beds, sports pitches, and parklands into allotments to produce food.

parklands into allotments to produce food (Figure 17). As stated in an editorial in the *Evening Standard*, the main London newspaper:

> Every spare half acre from the Shetland to the Scillies must feel the shear of the spade ... this war may be won in the farms as well as on the battlefields

Prior to the war, Britain was importing around 55 million tonnes of food each year, but during the war, German U-boats that stalked the seas cut off this supply. As a result, the Ministry of Food set up the 'Dig for Victory' campaign to feed the nation. It was a tremendous success: some 1.4 million people dug up their backyards and lawns to grow vegetables, and even the formal Royal Parks of London were turned to the spade. Allotments sprung up all over the country, often in the most unexpected places, such as at Manchester Ringway, which is now Manchester International Airport, and in exclusive polo grounds and bombed-out buildings in London. From the start of the war to 1943, the number of allotments in Britain rose sharply from just over 800,000 to 1.6 million, and the number of private gardens used for growing vegetables rose from 3 to 5 million. More than 1 million tonnes of vegetables were produced annually on the home front. Similar campaigns also happened in the United States during the Second World War. Here, the 'Victory Garden' movement encouraged civilians living in major cities and towns to dig up their gardens, yards, and parks to grow food, thereby supporting the war effort. Reports suggest that by the end of the war, there were more than 20 million victory gardens supplying 40 per cent of the produce consumed in America.[151]

Soil was central to these campaigns. Publicity images that supported the campaigns in Britain and America portrayed people turning to the land, digging the soil, using spades not ships, and sowing seeds, and newspapers and radios broadcast the call for civilians to cultivate the soil and grow their own. In Britain, 'Dig For Victory' exhibitions were held across the country, gardening became part of the school curriculum, and monthly *Allotment and Garden Guides* were distributed with tips on how to improve and get the best out of soil, such as: 'Never work the soil when it is too wet and sticky...seeds sown in cold, wet soil will rot instead of germinating.' In the United States, civilians were advised to choose their plots wisely and shown how to prepare the soil, but they didn't need chemical soil analysis because 'if it grows a fine crop of flowers or weeds, it's soil'. Across the country, Victory Garden Advisory Committees formed, people streamed to classes to become better gardeners, and conferences were held to discuss how best to garden soil. In Oregon, for example, the Oregon Victory Garden Advisory Committee formed in 1942, with the goal of 59,000 Victory Gardens in the state by the end of that year; by May 1943, children alone in the Portland area were cultivating over 100 acres of land and several classes were under way in the city to teach civilians how to get the best from their garden.[152]

Soil improvement to redress food shortages was also central to post-war agricultural reform. After the Second World War major efforts were made by governments of many counties to increase food production, and these efforts paved the way for continuous agricultural production since that time. This included the promotion of fertilizer and lime use to boost soil fertility, and the expansion of farming into new areas where infertile soils hampered yields. In Britain, for example, the Hill Farming Act was introduced in 1946 to provide grants to farmers to rehabilitate infertile soils in the hills where traditional, low-intensity sheep farming reigned. Farmers used these grants to fund the drainage of wet soils and the application of lime and fertilizers to land to boost soil fertility, grass growth, and production of meat.

In the US, bigger crop yields on fewer acres were the key to prosperity after the Second World War, and building up the fertility level of soil was central to this. Soils of the Corn Belt, for example, had been farmed more intensively during the Second World War with soil-depleting crops, such as corn and soybeans, than at any other period in their history. As a result, soils of the Corn Belt were depleted in organic matter and nutrients, and one of the major challenges faced by farmers was bringing about improvements in soil management to boost soil fertility and maintain yields. The Middle West Soil Improvement Committee,[153] for example, urged farmers to prepare themselves to meet the competitive conditions of post-war years by building up the fertility level of their soil, stating:

> The farmer who looks ahead now will have his soil in such a shape that high crop yields will cut his production costs to the point where he can withstand a period of lower prices.

When I started writing this chapter, I had no idea of the scale of impact that soil has had on war, and also how war has left its mark on soil. The sheer scale of these impacts is astonishing: artillery bombardments of unimaginable scale left soils pulverized during the First World War; the use of chemicals in Vietnam has left a deadly soil fingerprint that remains even today; and the nuclear age has left its signature on soils across the world. Moreover, war has left its fingerprint on soil in ways that I haven't even mentioned, such as the contamination of soil with oil during the Gulf War, and with heavy metals from fragments of shells during the First World War. Indeed, given the many and varied ways that warfare can impact soil, it has been argued that warfare is one of the most dramatic, but underappreciated ways by which humans can affect soil. Given how much of the Earth's soil has been affected by war, and how much is currently involved in war, there is little doubt that warfare should be recognized as a major factor shaping the soil.

6

●●●●●

Soil and Climate Change

It is commonly observed, that when two Englishmen meet, their first
talk is of the weather... *Samuel Johnson*

The world's climate is changing. Not only is it getting warmer, but
also there are more extreme weather events, such as droughts,
storms, and catastrophic floods. Humans are undoubtedly the
cause of this change in climate, through the burning of fossil
fuels, intensive farming, deforestation, and many other aspects of
our industrious lives that increase the emission of greenhouse
gases—carbon dioxide, methane, and nitrous oxide—to the atmosphere. In fact, over the past fifty years or so there has been an
unprecedented increase in the release of greenhouse gases to the
atmosphere, and, unless measures are put in place to cap emissions,
this trend is likely to continue.

So what have soils got to do with climate change? Put simply,
soils play a pivotal role because they act as both a source and sink
for greenhouse gases, and any disruption of this balance will affect
the concentration of these gases in the atmosphere and hence the
global climate, potentially making the situation either better or

worse. Perhaps the most powerful illustration of this concerns the carbon cycle. Soil is the Earth's third largest carbon store, next to the oceans and deep deposits of fossil fuels, and together with vegetation it contains at least three times more carbon than the atmosphere. Many worry that climate change will destabilize these carbon stores by stimulating the soil organisms that break down soil organic matter, releasing vast quantities of carbon dioxide to the atmosphere. This could shift soils from being sinks to sources of this greenhouse gas, thereby accelerating climate change. Scientists call this carbon-cycle feedback, and we will revisit it later.

THE GREENHOUSE GASES

Let's begin with the main actors of climate change, the greenhouse gases. The most abundant and well-known greenhouse gas is carbon dioxide. This gas is taken up from the atmosphere by plants through the process of photosynthesis, which occurs in the presence of light. Plants retain most of the carbon they take up and use it to grow and sustain their metabolism, but they also release a portion back to the atmosphere as carbon dioxide through respiration from both their shoots and roots. The main route by which carbon enters the soil is through the death of plants, and the addition to soil of roots, root exudates, and the dead aboveground plant remains. In fact, an astonishing 90 per cent of all plant material produced on Earth eventually ends up in the soil as dead organic matter, which is mostly made up of carbon. However, most of this organic matter doesn't stay in soil for long as the myriad microbes and animals that live there, the decomposers, break it down and use it for their growth. The activities of these decomposers, along with respiration of plant roots, yield vast amounts of carbon dioxide that is released back to the atmosphere. However, not all of this organic matter is broken down by soil organisms and released back to the atmosphere; especially in cold and wet places where organic matter decay is retarded, large amounts of organic matter can build up in soil, forming a big terrestrial carbon sink.

The next greenhouse gas, which is also a major player in the carbon cycle, is methane, which is an even more important greenhouse gas, with each molecule being about twenty-five times more powerful than carbon dioxide. Natural emissions of methane to the atmosphere result mostly from a process called methanogenesis, which is carried out by a very specialized group of microorganisms called archaea. These microorganisms thrive in very wet, oxygen-poor environments, such as wetlands, with waterlogged soils, oceans, animal rumens, and termite guts. As a buffer to the huge amounts of methane that are produced by archaeans, another group of microorganisms, the methanotrophic bacteria, consume a large portion of the methane produced in soils before it escapes to the atmosphere. Unlike methanogenesis, this is mainly an aerobic process, meaning that the microorganisms involved need oxygen to carry it out. Importantly, these microorganisms also consume methane from the atmosphere, and in doing so yield carbon dioxide which then enters the carbon cycle.

The final, but most potent greenhouse gas is nitrous oxide, each molecule being almost 300 times more powerful than carbon dioxide in its greenhouse warming power. There are many sources of nitrous oxide, but the most important is again the soil. Here, specialist groups of microorganisms called nitrifiers and denitrifiers carry out a range of transformations of nitrogen that produce nitrous oxide, which then escapes to the atmosphere. These transformations of nitrogen happen naturally, but one of the main sources of nitrous oxide emissions from soil is the use of inorganic fertilizers and manures. The issue here is that while plants take up a large amount of the nitrogen that is added to soil in fertilizers, not all of it is used, and that which isn't used by plants is converted into nitrate by nitrifying bacteria and archaea. Then, if conditions in soil become wet such that oxygen is lacking, part of this nitrate can be converted by denitrifying bacteria into a mixture of nitrogen gas and nitrous oxide, which then escapes to the atmosphere.

This process of denitrification is especially prevalent in soils that are rich in nitrate and when oxygen levels are low, such as in heavily fertilized grasslands in high rainfall regions, such as in western Britain. Here, the combination of high rainfall and heavy clay soils, coupled with high nitrate from fertilizers, creates ideal conditions for the production of nitrous oxide. Nitrous oxide is also produced as a by-product from the aerobic process of nitrification. This is the process through which soil microorganisms convert ammonium, derived from fertilizers or the breakdown of manures and soil organic matter, first to nitrite and then nitrate. Small quantities of nitrous oxide can then be formed and, in some parts of the world, this is a more important source of nitrous oxide than denitrification under wet and anaerobic conditions.

SOIL AS A CARBON SINK

As we noted earlier, soil is the third largest pool of carbon on Earth, next to deep geological deposits and ocean carbon. A remarkable 2,500 billion tonnes of carbon is held in the world's soils. The bulk of this carbon, around 1,550 billion tonnes, resides at the soil surface as organic matter that has built up over hundreds if not thousands of years. This organic matter is mostly made up of dead plant and animal material, along with substances synthesized by plants and microorganisms, and living micro-organisms and animals. The remaining carbon is inorganic, in the form of carbonates, such as calcite, dolomite, and gypsum. Soils vary tremendously in their inorganic carbon content, but it is especially high in arid and semi-arid environments, and on soils formed on chalk, such as those of the chalk downs of south-east England, or the Champagne region of France, where large amounts of calcium carbonate are found.

Peats are the most carbon-rich of soils, holding about one-third of the world's soil carbon stock. Very close to where I live there is a long belt of hills called the Pennines, which stretch from the Peak District in the south, just to the east of Manchester, northward into southern Scotland. These hills, which form the spine of northern England, are blanketed by extensive peatlands formed under the cold and wet conditions that have prevailed there for thousands of years. In some places, these peats can reach an astonishing 3 metres in depth, made up of sodden organic matter sourced from the mosses and dwarf shrubs that have grown there for centuries (Figure 18). This is just one example of peatland in a small part of the British Isles, a minute fraction of the vast peatlands that cloak the tundra and boreal zone, and the tropical peatlands that cover large areas of Africa and South East Asia; collectively, these peatlands make up a vast global carbon store.

Forests also have major soil carbon stores. I will discuss tropical forests later, but I first want to mention a recent discovery about soil carbon in boreal forests. These forests cover around a tenth of the Earth's land surface and store about a sixth of its total soil carbon pool. The conventional view is that this carbon mostly

FIGURE 18 Blanket peat of the Pennines, northern England. Here, deep peats have developed at high altitudes where high rainfall and low temperatures combine to cause excessive soil wetness that retards decomposition. Peatlands of this kind are a major store (c.30 per cent) of global terrestrial carbon, and the majority of the UK's terrestrial C is stored in the peat soils of northern Britain.

comes from dead needles and leaves of plants, which fall to the ground where they form a distinct humus layer on the soil surface. While dead needles and leaves certainly contribute to carbon stocks in forest soils, a group of Swedish scientists recently discovered that the majority of carbon in boreal forest soils comes from a completely different source: the remains of roots and root-associated mycorrhizal fungi.[154] It appears that trees direct much of their carbon to roots, and it is then transferred to the dense networks of mycorrhizal fungi that pack these soils. These fungi then incorporate some of this carbon into their mycelium, but when they die, their residues are slowly converted into organic material. What this means is that soil carbon doesn't just come from aboveground as plant litter, but it also comes from the remains of roots and root-associated fungi.

Even agricultural soils contain large amounts of carbon. In England, for example, grasslands are thought to contain about a third of our national soil carbon stock.[155] We recently looked into this more deeply; after sampling soils from a wide range of grasslands across the length and breadth of England, we discovered that, on average, a hectare of English grassland contains just over 400 tonnes of soil carbon to a depth of 1 metre (Figure 19).[156] The amount obviously varies from grassland to grassland, and the most carbon-rich of soils are those of wet grasslands, such as found in the Somerset Levels, which can hold upwards of 600 tonnes of carbon per hectare to 1-metre depth. These are considerable amounts of carbon, and much greater than previously thought. Another surprising finding of this survey was that the

FIGURE 19 Soil scientists have discovered considerable amounts of carbon stored deep in the soil profile.

way grasslands have been managed over the years has left a major fingerprint on soil carbon stocks. Grasslands managed in a traditional way, with low numbers of grazing animals and the occasional dressing of farmyard manure, were much richer in carbon than those that had been managed more intensively, with heavy dressings of fertilizers and high livestock stocking rates. In other words, what this result showed was that intensification of grassland management, which has occurred across many parts of Europe, has caused a loss of soil carbon.

Probably the worst thing for soil carbon is the conversion of natural grasslands and forests to arable agriculture. Not only does this reduce the amount of carbon entering soil, due to the loss of natural vegetation, which continuously feeds the soil with organic matter, but also tillage of soil causes dramatic losses of carbon. Tillage breaks up and mixes the soil, speeding up the breakdown of plant material. As a result, soils that are ploughed regularly and sown with crops tend to have lower stocks of soil carbon than those of grasslands or forests. Indeed, a large fraction of agricultural land globally has been degraded by over-cultivation and grazing, and this has caused major losses of soil organic matter, and hence carbon, in soil. In an attempt to overcome this, no-tillage agriculture is promoted in many parts of the world. As we saw earlier, this means that the soil isn't ploughed up annually. Instead, the residues of crops and roots are left behind to decompose in soil, leading to a build-up of organic matter and diversity of soil life in the surface soil. This practice of no-tillage farming is now used in many parts of the world to promote soil quality, through building up soil carbon reserves and reducing soil loss by erosion. But the downside is, as we noted, that lots of herbicides are needed to control weeds, and in some cases, no-tillage farming can boost emissions of nitrous oxide from soil.

Before leaving the topic of soil carbon stocks, I want to mention one sizeable pool that is often ignored: the carbon contained in soils of cities and towns. I talked about urban soils in Chapter 4, but it is worth mentioning again here that soils of gardens, parks, sports grounds, road verges, and allotments, contain sizeable amounts of carbon, and all this carbon contributes to the global carbon pool. It is not known how much carbon is locked up in urban soils because most is sealed beneath layers of tarmac, bricks, or concrete. But a recent study by a group of scientists from the universities of Sheffield, Kent, and Exeter discovered something quite startling:[157] that the amount of carbon contained in urban soils can be much greater than that found in nearby agricultural land. They came to this conclusion by sampling soils from a range of locations in Leicester, a midsized city in the English Midlands, and also discovered that the amount of soil carbon beneath buildings and other sealed surfaces, such as roads, pavements, and residential driveways, was the same as in soil under urban green spaces, such as parks and gardens. Although based on a single English city, these findings reveal that cities actually represent a sizeable, but mostly hidden, stock of soil carbon.

CLIMATE CHANGE AND SOIL CARBON CYCLING

Many scientists are worried that climate change will threaten carbon stocks in soil. The worry is that warming of soils will stimulate the activity of the microbes that break down organic matter, and that this will increase the release of carbon from soil to atmosphere, thereby further accelerating climate change. This appears to be happening already in some parts of the world. Research done in Abisko, a remote place in northern Sweden, showed that even a small increase in air temperature of around 1 °C can stimulate soil microbes, thereby boosting decomposition and carbon loss from these tundra soils.[158] Researchers monitoring Alaskan tundra also found that permafrost thaw, which is speeding up with climate change, has ramped up microbial activity and carbon loss from soil, despite increased carbon capture from the air due to greater plant growth.[159] This is especially worrying because tundra soils are among the world's largest carbon stores, and about a quarter of the Earth's permafrost is predicted to thaw by the end of this century.[160]

Another striking outcome of climate change is its effect on mushrooms. Alan Gange, an expert in soil fungi from Royal Holloway, London, looked back at records over the last fifty years of the first fruiting date of 315 species of mushrooms in southern England.[161] What he and his colleagues found was remarkable: the average first fruiting date of mushrooms is now earlier, while the last fruiting date is later, than was the case fifty years ago. In sum, they discovered that the overall fruiting period has more than doubled since 1950s, from around thirty-three to seventy-five days, and many fungi have switched to fruiting twice a year, once in spring and again in autumn. Similar patterns have been found in other parts of the world. A group of Norwegian scientists looked at herbarium records over the period 1940–2006, and also found that the time of fruiting of mushrooms in Norway has changed considerably, being much later in autumn.[162] As in Gange's study, these changes were put down to climate change, and especially to an extended growing season. You might be wondering why this is important for soil carbon. The reason is simple: fungi are major players in the breakdown of dead plant remains, so any increase in the length of time that they are active will boost organic-matter decay and the cycling of carbon in soil.

It would be wrong to give the impression that the effect of warming on soils is straightforward and that all soils will lose carbon as they warm. This is not the case: some soils do respond to warming by losing carbon, such as the already-mentioned peat soils of the Arctic, but others do not. For instance, researchers studying tall-grass prairie on the US Great Plains in Central Oklahoma found that continuous experimental warming by 2 °C since 1999 didn't affect the total amount of carbon in soil; it simply didn't change, despite big changes in vegetation and the composition of the microbial community in soil.[163] Also, where warming does result in soil carbon loss, the effect sometimes diminishes with time as soil microbes adapt to warmer temperatures. With time, the carbon that is lost from soil because of warming could also be compensated for by increased storage of carbon in plants. As an example, experiments done in a deciduous forest in Massachusetts, USA, showed that although soil warming caused carbon loss from soil, this was compensated for by carbon gains in the woody tissue of trees.[164] This was because warming also increased the cycling and availability of nitrogen in soil, which supported the increase in tree growth and therefore plant carbon storage.

Some of the most dramatic effects of climate change on soils result from extreme weather events, such as droughts, storms, flash floods, and unusually cold spells, which are all becoming increasingly common. As I write, we are experiencing in England one of the coldest springs on record, and last year was among the wettest, with soils being waterlogged for most of the year. Similar stories of extreme weather come from all over the world, and while climate variation is the norm, the increase in its intensity is almost certainly due to climate change. Extreme weather events, such as drought and freezing, can place considerable stress on the microbes and animals that live in soil, often resulting in their death. But when soils re-wet or thaw after prolonged drought or freezing there is often a tremendous flush of microbial activity that yields a pulse of carbon dioxide and nutrient release from soil. This flush is caused by the breakdown of organic matter that was previously inaccessible to microbes, such as that trapped in ice, and the degradation of dead microbial cells; it is referred to as the 'Birch effect', after the British scientist H. F. Birch who discovered, while working in Africa during the 1950s and 1960s, that cycles of drying and wetting substantially boosted the release of nutrients and carbon from soil. The scale of these effects can be enormous: in Mediterranean and savannah ecosystems which experience lengthy periods of drought, the pulse of carbon dioxide following the first rainfall after a dry summer can substantially reduce the amount of carbon that these ecosystems gain each year. And in the Arctic, the thawing of frozen soils after the long winter leads to a burst of carbon dioxide and methane release from soil.

The Amazon is worth a special mention because of the enormous impact recent droughts have had on carbon release to the atmosphere. The Amazon rainforest covers a vast area of the world's land surface. In a normal year it absorbs more than 1 billion tonnes of carbon dioxide from the atmosphere, which counters emissions caused by deforestation, logging, and fire. But in 2005 and 2010 the Amazon was struck by unusually severe droughts, which killed trees on such a scale that the forest stopped absorbing carbon dioxide from the atmosphere. The problems didn't end there. As the dead trees decomposed, they released even more carbon dioxide to the atmosphere. A study by a team of British and Brazilian scientists, revealed the enormous scale of this carbon release.[165] They calculated that the 2005 drought, which was described as a 1 in 100 year event, caused 5 billion tonnes of carbon dioxide to be released into the atmosphere, and they suggested that emissions resulting from the 2010 drought would be even bigger. To put this into context, these two extreme droughts, which unusually occurred within a decade, could offset the entire amount of carbon absorbed by the intact Amazon forest over that time. More worryingly, they predicted that if such droughts occur more often in the future, the Amazon could shift from being a global sink for carbon, to a major source of carbon dioxide that could accelerate climate change. Similar stories are emerging elsewhere. For example, the record-breaking heat wave that hit Europe in 2003 led to the release of more carbon dioxide than is normally locked up over four years. Moreover, the probability of similar heat waves hitting Europe is expected to surge over coming decades with large impacts on the ability of land to sink carbon.

Another consequence of climate change is reduced snow cover in the high mountains. Snow cover in the mountains varies greatly from year to year, but as any skier will know, as temperatures continue to rise, many ski resorts, especially those at lower elevations, will no longer be viable because of reduced snow. Studies in the European Alps reveal a general trend of declining snow cover over the twentieth century, and some predict that snow cover at 2,000 metres' elevation will reduce by half, which equates to around two months' duration, by the end of the century.[166] This will put many European ski resorts at risk, especially those below 2,000 metres' elevation. The picture is no better in other countries such as the United States, where snow cover is projected to reduce substantially over the twenty-first century, and in some ski areas it could drop to zero.[167]

While devastating for the winter snow industry, the impacts of reduced snow cover on soils are also predicted to be profound. This is because snow acts as an insulator of soils in winter, and reductions in snow cover increase soil freezing, which can strongly affect the soil, leading to increased activity of microbial processes and greenhouse gas emissions from soil after thawing. As with drying and wetting cycles, freeze-thaw kills some organisms and renders a part of non-living soil organic matter more decomposable, resulting in a flush of nutrients and carbon dioxide release from soil. Deep-freezing can also cause roots to die, negatively affecting the growth of plants. These effects are poorly understood, but they greatly affect soil carbon storage and greenhouse gas emissions in alpine regions of the world.

Climate change can also affect soil carbon through its impact on soil erosion, which not only has devastating effects on the ability of land to support crops, but also causes loss of carbon from soil. In Chapter 3, I discussed how vast amounts of soil are lost from croplands worldwide due to soil erosion, and how much of this soil ends up in rivers and oceans, along with the carbon that it contained. Climate change is likely to exacerbate these problems, with drier soils being more susceptible to the erosive forces of wind and rain, and more intense rainstorms increasing soil erosion and carbon loss from soil. The picture gets even more complicated when you take into account the shifts in land use that are likely to occur as a result of a changing climate. In many parts of the world, farmers might have to grow new crops, or even completely change the way they farm their land, and this could have major effects on rates of soil erosion and carbon loss from soil.[168] It is very difficult to predict how farming will change in the future under climate change, but there is concern that the effects of changing land use on soil erosion could be considerable, and more important than direct effects of climate change alone.[169]

CLIMATE, PLANTS, AND SOILS

Some of the strongest impacts of climate change on soils cascade from plants. This can be through rapid changes in the growth and physiology of plants, for example in response to higher concentrations of carbon dioxide in the atmosphere. Or, it can be through climate-change-driven alterations in the diversity or make-up of plant communities that occur over tens or even hundreds of years. What both of these have in common is that they change both the amount and type of carbon that goes into soil as exudates, which leak out of roots, and plant litter. These inputs go on to impact soil organisms and the breakdown and recycling of carbon.

As atmospheric carbon dioxide concentrations increase, plants grow faster and bigger, storing more carbon in their biomass and helping to mitigate climate change. But scientists have also discovered that when plants take up more carbon dioxide by photosynthesis, they also excrete more carbon from their roots into soil, where it boosts the growth and activity of microbes. A study in a pine forest in North Carolina, USA,[170] showed that this cascade of carbon from roots to soil can have major effects on the carbon budget of the ecosystem.[171] They found that twelve years of elevated atmospheric carbon dioxide not only increased the flux of carbon to roots, which boosted the activity of microbes in soil, but also caused greater breakdown of organic matter, which led to carbon loss from soil. But that wasn't the end of the story. Because this process also stimulated the availability of nitrogen in soil and its uptake by the trees, it set in motion a positive feedback loop that sustained increased rates of tree growth under elevated carbon dioxide, which locked up even more carbon in the trees. So, the overall consequence of this feedback loop was an increase in the amount of carbon stored in trees, but at the expense of carbon in soil.

As with many things that concern the soil, things are not so simple. Other studies show that boosting carbon supply to soil from roots under elevated carbon dioxide causes the locking-up of nitrogen by soil microbes. This process, which is called nitrogen immobilization, has the opposite effect to what was found in the North Carolina pine forest. It reduces nitrogen supply to plants, which curtails plant growth under high carbon dioxide, thereby reducing the amount of extra carbon locked up in trees. It is unclear why elevated carbon dioxide boosts soil nitrogen avail-ability in some cases but not in others. However, it is probably due to differences in the chemical make-up of the exudates and plant litter that enters soil. When they are rich in nitrogen, soil microbes use carbon contained with the exudates and litter to sustain their growth, but they release into soil the excess nitrogen that they don't need, thereby making more available for plants, or in agri-cultural soils, more vulnerable to leaching. When exudates and plant litter are low in nitrogen, however, the microbes become nitrogen-starved and soak up any nitrogen from the soil, therefore making it less available to plants. This often occurs because while elevated carbon dioxide boosts the amount of carbon supplied to plants, it doesn't provide extra nitrogen, so plant tissues become depleted in nitrogen relative to carbon. So, unless extra nitrogen is supplied to plants, for example from fertilizers, the exudates and litter entering soil in a high carbon dioxide world are often poor in nitrogen.

As well as affecting the growth of individual plants, climate change can cause major shifts in the diversity and structure of natural plant communities. Recent changes in rainfall, for example, have caused major shifts in the make-up of vegetation in tropical rainforest and African savannah, and warming is leading to rapid losses of plant diversity in salt marshes, and even to increased tree mortality in some parts of the world. Such shifts in the natural vegetation affect not only the amount and chemical composition of organic matter entering soil, but also the soil's physical environment because of changes in root growth that affect the structure and moisture content of the soil; for these reasons, they can have considerable knock-on effects for soils and climate change.

One of the most dramatic examples of climate change affecting vegetation and soils comes from the Arctic. The Arctic is a cold and inhospitable place that is extremely vulnerable to climatic warming. A much-written-about consequence of this warming is the northwards expansion of shrubs, such as dwarf birch (*Betula nana*), or the greening of the Arctic.[172] The reason why this is important for soil carbon storage is because shrubs, like dwarf birch, produce leaf litter and woody material which decomposes much slower than the dead litter of the grasses and herbs that they replace. Because of this, researchers believe that the expansion of shrubs across the Arctic could lead to a rapid build-up of litter on the soil surface, which over the years could ramp up carbon content of soil. Whether this build-up of carbon is enough to counter the increase in organic matter decomposition and soil carbon loss that results from warming is not known. This example shows that shifts in vegetation resulting from climate change could dampen effects of warming on carbon loss from soil. But, more generally, it also shows that the impact of climate change on carbon sequestration depends not just on how soils respond, but also on how vegetation is affected, and the balance between the two.

The picture becomes even more complicated when grazing animals come into the picture. A graphic example of this comes from a study by scientists at Penn State University, USA, and the University of Aarhus, Denmark.[173] They set up an experiment in West Greenland to see how muskoxen and caribou changed the way that vegetation responded to climate warming. What they found was quite remarkable: when these animals were absent, warming boosted the growth of willow and birch, which shifted the vegetation from being dominated by grasses to being covered in these woody shrubs. Because of this more carbon was stored in the vegetation and most likely the soil, given that woody shrubs cause a build-up of woody litter on the soil surface. But when muskoxen and caribou were browsing on trees, their response to warming was kept in check, so the vegetation remained dominated by grasses. This finding is important because it shows that natural grazers of the Arctic substantially alter carbon storage by Arctic vegetation and presumably also soil. Moreover, it is likely that grazing animals have similar effects in other parts of the word, completely changing the way ecosystems respond to climate change.

Grazing animals also affect greenhouse gas emissions from soil. Most interest in this area revolves around grassland farming and the fact that high stocking rates of cattle or sheep can increase greenhouse gas emissions, especially of nitrous oxide. This is because grazing generally increases nitrogen cycling, and faster cycling of nitrogen causes higher emissions of nitrous oxide from soil. But a study by Benjamin Wolf of the Institute for Meteorology and Climate Research, Germany, and his colleagues, showed that the reverse can be true: in some places, grazing animals can actually reduce emissions of greenhouse gases.[174] They looked at nitrous oxide emissions across a range of semi-arid grassland sites in Inner Mongolia, and found that grazing decreased emissions of nitrous oxide during the spring, which is the time when most annual emissions occur. The springtime pulse in gas emissions in this region occurs because the subsurface soil is still frozen at that time, and melting snow saturates the surface soil layers, creating anaerobic conditions that are especially conducive to denitrification and therefore nitrous oxide production.

What Wolf and his colleagues discovered was that grazing reduced plant height and biomass, and because of this, less snow was trapped and hence soil was less sodden at snowmelt. Because snow also insulates the soil over the winter, the drop in snow cover in grazed grasslands made the soils colder. Together, the lower amounts of water coming from melting snow and the colder soil conditions of grazed sites acted to inhibit the activity of microbes involved in denitrification, thereby reducing nitrous oxide emissions from soil. The scale of this effect was enormous, with emissions of nitrous oxide from sites without grazing animals being more than double than from those that were grazed. These findings are important because they show that studies that look only at greenhouse gas emissions during the growing season could lead to faulty estimates, given that most nitrous oxide emissions from these grasslands were in spring, well before the main growing season had begun. They also show that the way that we manage the land, in this case by grazing animals, can greatly influence greenhouse gas emissions from soil.

Climate change can impact soil carbon cycling in the most unexpected ways. As I have already hinted, climate change is causing major shifts in species ranges, and many species are expanding their ranges upwards to higher elevations and northwards to the poles. But it is not just plants that are moving; animals are also expanding their ranges, both aboveground and belowground, and this is disrupting carbon cycles in soil, as work by Carol Melody and Olaf Schmidt from Dublin has shown. In their studies of the northward march into Ireland of the Mediterranean earthworm *Prosellodrilus amplisetosus*, which has been put down to rising soil temperatures, they discovered that not only has this earthworm become the dominant worm species in the Irish sites they sampled, but it is also feeding on types of soil carbon that are inaccessible to the resident species.[175] Because of this, they warned that the northward movement of this earthworm could reduce carbon stocks in Irish soils.

Another example concerns the recent outbreak of the mountain pine beetle *Dendroctonus ponderosae* that has been devastating forests across western North America. Pine beetle outbreaks have been put down to many factors, such as the recent increase in area of mature pine stands. But it is also caused by climate change, which has driven the outbreak northwards and into higher elevation forests. So, what has this got to do with the carbon cycle? Werner Kurz and his team from the Canadian Forest Service asked this very question.[176] Using computer models, they estimated that the recent outbreak of the mountain pine beetle in British Columbia had actually converted the forest from a small net carbon sink to a large net carbon source. This was put down to a reduction in carbon uptake caused by widespread tree mortality, combined with an increase in carbon dioxide emission from the rotting of dead trees. The scale of impact on carbon emissions is enormous, being equivalent to five years of greenhouse gas emissions from Canada's transportation sector (i.e., 200 megatonnes of carbon dioxide equivalents in 2005). As the author Werner Kutz commented in an interview about his paper: 'Those are very big numbers brought on by a very small insect.'

STORING CARBON

A major challenge facing the world is how to mitigate climate change. Many ways of dealing with this have been put forward, such as switching from burning fossil fuels to using non-fossil fuel sources of energy, including wind power, nuclear energy, and geothermal sources. More elaborate schemes involve injecting carbon dioxide deep into the oceans or geological strata, and even using a cloud of small spacecraft to act as 'sunshades' in space, reflecting solar radiation. So where does soil come in? Put simply, soil is one of the world's largest pools of carbon, so by increasing its size further, it should be possible to draw down carbon dioxide from the atmosphere, thereby mitigating climate change. Although scientists are divided on the benefits of soil carbon sequestration for combating climate change,[177] it has been proposed that by changing the way we manage the world's agricultural and degraded soils, we could boost soil carbon pools by 0.4 to 1.2 billion tonnes a year, which is equivalent to 5 to 15 per cent of the global fossil-fuel emissions.[178] As I discussed in Chapter 3, this building of soil organic matter in agricultural and degraded soils also reaps benefits for soil fertility, including improved soil structure and water-holding capacity, and increased storage and retention of nutrients such as nitrogen and phosphorus. In other words, increasing soil carbon content doesn't just benefit climate mitigation; it is also central to the fertility of soil.

The amount of carbon in soil is determined by how much goes in from plants and how much goes out through decomposition of organic matter, burning, and soil erosion. To increase soil carbon content you therefore need to put more carbon into soil and at the same time reduce how much comes out. Many of the world's soils are depleted in carbon as a result of the burden of decades of heavy cultivation and overgrazing. And many ways have been suggested to reverse this process by changing the way we grow crops or by slowing down the decomposition of dead plant matter in soil. These include the use of no-tillage farming, which cuts out soil disturbance by tillage and the breakdown of crop residues, the conversion of arable land to perennial grassland, which causes a build-up of organic matter at the soil surface because of increased root growth and the lack of tillage, and using cover crops in rotations, which increase the input of organic matter input to soil. Of course, taking land out of arable agriculture to lock up more carbon in soil, while good for combating climate change, works against another major global challenge: producing enough food to feed a growing population.

The scale of the benefits for carbon storage of changing soil management can be considerable. As touched on above, the conversion of large tracts of land to no-tillage agriculture in many parts of the world, such as Americas and Australia, has been proposed as a way of boosting soil carbon, with added benefits for soil fertility. The restoration of woodland and grassland on degraded arable land had also been put forward as a way to build up carbon stocks in soil, although the global trend is in the opposite direction: forests and savannah grasslands are being converted to arable cropping. Over the past few decades in France, for example, vast areas of grassland have shifted to arable land, causing major drops in soil carbon. Scientists have suggested that bringing back just one-half of this arable land to permanent grassland, which would amount to around 90,000 hectares of land, would increase France's soil carbon stock by around 16 million tonnes.[179] To put this into perspective, this quantity of carbon is equivalent to around 10 per cent of the annual carbon dioxide emissions from fossil fuels in the whole of France.

While there can be large gains for soil carbon storage through changes in land use, there are also costs; some farming practices that build soil carbon can also promote emissions of nitrous oxide from soil. Given the potency of nitrous oxide as a greenhouse gas, this can cancel out any benefits for climate mitigation of the increased soil carbon stock. This dilemma is especially the case for no-tillage farming; while it can build up soil carbon, it can also boost emissions of nitrous oxide, thereby potentially negating benefits gained from increased carbon storage. Johan Six, an expert in soil carbon based in Zurich, and his colleagues looked into this in more detail.[180] They discovered that in many situations, newly converted no-tillage systems had a higher global warming potential than did conventional systems, mostly because of increased nitrous oxide emissions. This finding led Johan and his colleagues to argue that the promotion of no-tillage for carbon storage is naïve since its effect on the net greenhouse balance is very variable and unclear. More recent assessments support this view, suggesting that while no-tillage farming brings benefits for soil fertility, its impact on climate mitigation is overstated.[181] Other soil scientists have looked more deeply into this issue, and it appears that the way that no-tillage farming affects the emission of nitrous oxide depends on the nature of the soil, and especially how wet it is. A study by Phillipe Rochette of Canada, for example, revealed that no-tillage farming generally stimulates nitrous oxide emissions in poorly drained soils, but has little effect in soils with good and medium drainage.[182] In other words, it seems that increased emissions of nitrous oxide following conversion to non-tillage agriculture might only be a problem in poorly drained soils and in regions of the world with high rainfall.

Another way of boosting soil carbon is through changing the way grasslands are managed. Grasslands play a special role in carbon storage because they cover such a vast proportion of the Earth's land surface. Soils of grasslands are also rich in carbon, especially at the surface where dense root systems help to bind soil and build up carbon. Added to this, grasslands are also havens for biodiversity, supporting large numbers of plant and invertebrate species. As I mentioned earlier in this chapter, in many parts of the world the carbon content of many grassland soils, and also their biodiversity, has been severely depleted because of decades of intensive farming and overgrazing, often causing drastic soil degradation. Because of this, scientists and land managers have been trying to work out how best to restore the biodiversity to these grasslands, whilst at the same time reaping rewards for soil fertility and carbon storage.

New research suggests that planting of grasslands with diverse plant mixtures might be an effective way of tackling this problem. In one study in the USA, scientists planted high-diversity mixtures of native grassland plants into soils that had been severely degraded by years of intensive farming. What they found was that these high-diversity mixtures not only boosted crop yields, but also reduced greenhouse gas emissions compared to single-species monocultures, and brought added benefits for wildlife conservation.[183] In another study in Minnesota, USA, carbon-poor soils planted with high-diversity mixtures of perennial plant species some twelve years earlier were found to store an astonishing 500 per cent more carbon than did single-species plots.[184] This remarkable effect was put down to the presence of certain grasses and legumes in the high-diversity mixtures, which worked together to boost root growth and carbon input to soil.

Legumes appear to play a special role in building carbon in soil. In an experiment set up in the Yorkshire Dales in northern England, we discovered, to our surprise, that by simply sowing the legume red clover (*Trifolium pratense*) into species-rich meadows, we boosted soil carbon by around 10 per cent in just over two years.[185] This high rate of carbon build-up was also linked to a drop in emissions of the greenhouse gas carbon dioxide and the structure of soil was also improved, being more porous and better drained. We don't know the reason for this rapid build-up in carbon, but the legume also reduced the activity of certain enzymes in soil that microorganisms produce to break down organic matter. Also, the improved soil structure meant that more carbon was locked away in soil aggregates, where it was protected from microbial attack. For these two reasons, we think the legume triggered the build-up of carbon in soil.

From all these studies, it is clear that roots play a crucial role in building carbon in soil. Not only do they act as the conduit for supplying carbon to soil from plants but, depending on root depth characteristics, they also have the potential to place the carbon deep in the soil profile where it is less vulnerable to the vagaries of climate change and also the plough. With this in mind, scientists have started to wonder whether it might be possible to actually breed crops that produce more and deeper roots, which may maximize carbon input to soil, and hence carbon storage underground. The argument for this is based on three main points.[186] First, agricultural soils worldwide are depleted in carbon, especially at depth, so there is substantial scope for increasing the amount of carbon stored in them. Second, roots, and especially their density and depth down the soil profile, are crucial for building soil carbon. Finally, plant scientists have known for many years that it is the genetic make-up of a plant that determines how deep and bushy its roots can become. All these points suggest that breeding programmes could be targeted at promoting root architectures that can sequester carbon in soil more effectively, whilst also capturing water and nutrients. This is a long way off, and there is a potential problem of the extra carbon going to roots reducing aboveground yields; but it might become a reality in the distant future.

The message of this chapter is simple: soils play a major role in the carbon cycle and in regulating climate change. Also, there is much that we can do to boost soil carbon storage and therefore mitigate climate change, with added benefits for the fertility of degraded soils. These benefits might be regional or even national, for example through changes in the way we manage farmland. Or they might be local, such as through planting certain plant species in gardens or fields to boost carbon supply to soil or reduce its loss. Having said this, there is still much to be learned about how climate change affects soils and the carbon cycle, and how crops and land management can be changed to speed up carbon sequestration in soil. But what is clear is that new research initiatives are needed to tackle the challenges for soil of balancing land management and climate mitigation as global warming takes its toll.

7

Soil and the Future

I bequeath myself to the dirt, to grow from the grass I love; if you want me again, look for me under your boot-soles.   *Walt Whitman*

If the importance of soil for human lives hasn't leapt out of the previous pages, this book has failed in its goal. Soil touches so many aspects of human life, often in ways of which we are not even aware. There are the obvious, such as when we dig up the soil to grow vegetables and flowers, or when a farmer takes a plough to a field. But there are also the less obvious, such as the role of soil in dampening climate change, filtering the water we drink, and breaking down and recycling the billions of tonnes of dead plant remains that annually fall to the ground. Soils have also played their role in warfare, thwarting military advances and providing underground shelter to those under attack. I could go on, but I think the message is clear: earth matters.

Looking to the future, a major challenge for humans will be how to deal with rapid soil change. I emphasized at the start of this book that the natural rate of soil formation is spectacularly slow; it takes literally thousands of years for a mature soil to develop.

But within just a few years, or decades, humans can completely transform the structure, chemistry, and biology of soils, often leading to their degradation. This degradation of soil can be catastrophic, for example when soils are over-cultivated or over-grazed, or when unstable hill slopes are deforested and left exposed to the erosive forces of wind and rain. Or it can be progressive, such as that caused by climate warming which, in some places, such as the Arctic, is gradually speeding up organic-matter decay and carbon dioxide release from soils. It can also be abrupt, such as when land is sealed by asphalt and concrete during the construction of expanding cities, or during war when major offensives obliterate the fabric of soil. As I stressed earlier in this book, the causes of soil degradation are complex: population growth, poverty, poor delivery of information to farmers, conflict, shortage of land, and climate change all play a role.

Not all soils are degraded. On the contrary, the Earth is still rich in natural soil variation, and in the same way that biologists celebrate the diversity of life, soil scientists revel in the diversity of soil. Also, many soils of agricultural landscapes are extremely fertile, as long as care is taken to replenish them with organic matter and nutrients, and to counter the degradative forces of compaction, acidification, and salinization of soil. And in cities, soils of allotments and gardens can be healthier than those of surrounding agricultural fields, as long as they are treated well. Soils can also be remarkably resilient, and while past soil damage can leave some soils compromised for many centuries, others recover relatively fast. Take the soils of the Western Front, which were completely obliterated during the First World War. Although it has taken more than a century, these soils are starting to show signs of recovery, with soil horizons beginning to show. And in some parts of the world, such as China's Loess Plateau, major multimillion-dollar initiatives have successfully curbed soil erosion and revitalized degraded soil, which has increased farmer incomes and lifted people out of poverty.[187]

But there is little scope for complacency. Vast areas of the Earth's land surface are in a state of degradation and unable to support healthy crops, and a new breed of studies is revealing the growing extent of the world's rare and endangered soils. An exhaustive assessment of the diversity of soils in China, for example, revealed eighty-eight endangered soils across the nation, and a further seventeen that have become extinct due to pressures of farming and urban development.[188] They also found another 230 types of 'new' soil, covering some 12 per cent of the nation's land surface, created by centuries of farming, which has completely transformed the natural soil. A similar picture emerges in the US where a landmark study of the nation's soil diversity revealed an astonishing 500 'endangered' soils, at risk from human activities such as farming and urbanization, and several that had already gone extinct.[189] These are just a couple of landmark studies, but what is worrying is that their findings likely mirror what is happening elsewhere: humans are causing not just the extinction of plant and animal species around the globe, but also the extinction of species of soil. And just as biologists are making major global efforts to protect biological diversity on Earth, soil scientists are now stepping up efforts to preserve the diversity of soil.[190]

Given the growing extent of soil degradation, a question that troubles many is whether the world's soils can support the growing demand for food. By 2050, the world's human population is predicted to increase by 9 billion, placing considerable demand on land to produce more food. And while the last fifty years have witnessed remarkable growth in the production of food, it has come at a considerable cost to the soil. In some parts of the world, yields have staggered or completely collapsed due to degradation of soil, and history tells us that continued mismanagement of the soil can have catastrophic consequences for mankind. The world is therefore presented with a major conundrum: food production needs to be increased, and distributed more equally, but it must be done sustainably, and in a way that counters negative impacts on soil and the many other benefits that it provides for humans. This also needs to be done at a time of rapid climate change, with rising temperatures and deepening droughts threatening food production and placing considerable pressure on soils, and growing pressures on land availability, for example from expanding cities and protected areas of land. All these factors create a daunting and complex challenge, which requires tackling many constraints on food production, not just the need for healthy soil. Moreover, it requires addressing not just technological and practical challenges for improving crop production and the management of soil, but also political and societal challenges to ensure food supplies across the world are effectively distributed and safeguarded in a sustainable way.

Much has been written elsewhere about dealing with this challenge of feeding the world in a sustainable way,[191] so I don't want to go into this here. Rather, I want to finish by considering, more generally, the future for soils, and especially the urgent need for increased awareness of their vital importance for human lives, not just for the production of food, but also in the many other ways that I have considered in this book. The more general question I want to depart with is: what can be done to improve the management of soil?

First, there is a need for greater awareness among society at large of the vital importance of soil for human lives. This was the main goal of this book, to awaken awareness of the many ways that soil has affected, and continues to affect, our lives. Many readers will already appreciate the value of soil, and soil stories are now being highlighted more in the media, but there is a fundamental need for educational programmes for all members of society to build knowledge of the importance of soil. From an early age, schoolchildren should be taught about the vital importance of soil and its role in sustaining life, and the scientific study of soil should be embedded in national school curriculums. There is also a need for more public events, in museums, galleries, and festivals, to celebrate soils, and for training of practitioners in the skills needed to understand the soil. Universities should also work with funding bodies to halt the decline in soil science, and to redress skills gaps in the application of soil science to relevant fields of agriculture, plant science, ecology, and environmental science. This is already happening in some parts of the world, such as the UK, where funding agencies have identified soil science as a 'most wanted' skills gap in urgent need of targeted training;[192] but more needs to be done.

Scientific understanding of soil also needs to advance. But this needs to be done in a holistic way, integrating new knowledge on the behaviour of soils into wider understanding of how to manage soils for growing food, protecting clean water, and mitigating climate change. There is also a need for a step change in our understanding of how soils in different parts of the world continue to work under the pressures of land use and climate change, and to devise ways of mitigating the forces of soil degradation, especially in the most vulnerable parts of the world. The shelves of university libraries are already replete with books on every imaginable aspect of the study of soil, and academic journals are crammed with innovative soil research; fascinating new discoveries about the wonder of soil emerge every day. But there are still many scientific discoveries to be made. And in order to tackle the big challenges facing soils, there is a need for more holistic studies of the soil that consider in unison their physics, chemistry, and biology. More-over, the big advances in understanding of soil are likely to come by working with scientists from other disciplines, such as plant science, microbiology, geology, and modelling of climate change, which will serve to create more holistic ways of managing soil.

In order to be of use, this growing scientific understanding needs to be translated into practice, so there is also a need for scientists to work with farmers, land managers and advisers, urban planners, and policymakers. Things are certainly changing on these fronts, and a number of new research programmes on soils in many parts of the world, including the most vulnerable such as Africa, South America, and China, have been announced, or are being planned for the future. But given the central role of soils in meeting this century's global challenges, and the strong link between soil fertility and human well-being, more funding is needed to support major national and international research efforts designed to deliver holistic solutions to the management of soils.

To protect soils and manage them sustainably, there is also need to embed soil issues in all future decisions on the use of land, whether that be for farming, forestry, flood prevention, the design of cities, industrial development, or the protection of land to conserve biodiversity. This is already starting to happen, and even within the time that I have been writing this book, new mandates for managing soils have been put in place.[193] Several of the United Nations (UN) Sustainable Development Goals for 2016 to 2030 relate to soils, and the UN's Intergovernmental Technical Panel on Soils' report[194] documents how soils are changing globally and the consequences of this for humanity. New international initiatives are also springing up to promote careful management of soil and the need for nations to implement monitoring programmes to document soil change. The Global Soils Partnership, for example, provides advice to governments and interested parties on global issues related to soil, and the Global Soil Biodiversity Initiative was established in 2011 to provide a platform for promoting the translation of scientific knowledge on the biodiversity of soil into environmental policy and sustainable management of land. These are just a few of the initiatives set up to raise awareness of the Earth's soil and its importance for humans.

Finally, attitudes to soil need to change. Soil needs to be considered as an investment to be protected and cared for, and as part of the support network for human life. All too often, soil has been ignored, or treated with complacency and little regard for its health. As we have seen, soil influences human lives in so many ways. It is a precious resource. And if it is not treated with respect, it will be gone for good. As many have said before, civilization has its roots in the soil, and without soil there will be no future life.

NOTES AND REFERENCES

1. Bierman, P. R., et al. (2014). Preservation of a preglacial landscape under the venter of the Greenland ice sheet. *Science* 344, 402–5.
2. Crowe, S. A., et al. (2013). Atmospheric oxygenation three billion years ago. *Nature* 501, 535–8.
3. Jenny, H. (1941). *Factors of Soil Formation*. New York: McGraw Hill Book Company.
4. Bardgett, R. D. (2005). *The Biology of Soil*. Oxford: Oxford University Press.
5. Jenny (1941). *Factors of Soil Formation*.
6. Jenny (1941). *Factors of Soil Formation*.
7. Vitousek, P. M., and Walker, L. R. (1989). Biological invasion by Myrica faya in Hawai'i: Plant demography, nitrogen fixation, ecosystem effects. *Ecological Monographs* 59, 247–65.
8. Stockmann, U., Minasny, B., and McBratney, A. B. (2014). How fast does soil grow? *Geoderma* 216, 48–61.
9. Walker, T. W., and Syers, J. K. (1976). The fate of phosphorus during pedogenesis. *Geoderma* 15, 1–19.
10. Walker and Syers (1976). The fate of phosphorus during pedogenesis.
11. Wardle, D. A., et al. (2004). Ecosystem properties and forest decline in contrasting long-term chronosequences. *Science* 305, 509–13.
12. Hall, S. J., et al. (2013). Legacies of prehistoric agricultural practices within plant and soil properties across an arid ecosystem. *Ecosystems* 16, 1273–93.
13. Dupouey, J. L., et al. (2002). Irreversible impact of past land use on forest soils and biodiversity. *Ecology* 83, 2978–84.
14. Glaser, B., and Birk, J. J. (2012). State of the scientific knowledge on properties and genesis of Anthropogenic Dark Earths in Central Amazonia (terra preta de Índio). *Geochimica et Cosmochimica Acta* 82, 39–51.
15. Turner, R. C., Rhodes, M., and Wild, J. P. (1991). The Roman body found on Grewelthorpe Moor in 1850: A reappraisal. *Britannia* 22, 191–201.
16. Wald, C. (2105). Forensic science: The soil sleuth. *Nature* 520, 422–4.
17. Decaëns, T., et al. (2006). The values of soil animals for conservation biology. *European Journal of Soil Biology* 42, 23–38.

18. Darwin, C. R. (1881). *The Formation of Vegetable Mould through the Action of Worms and Observations on Their Habits*. London: John Murray.
19. Bardgett, R. D., and van der Putten, W. H. (2014). Belowground biodiversity and ecosystem functioning. *Nature* 515, 505–11.
20. Kiers, E. T., et al. (2011). Reciprocal rewards stabilize cooperation in the mycorrhizal symbiosis. *Science* 333, 880–2.
21. Roesch, L. F., et al. (2007). Pyrosequencing enumerates and contrasts soil microbial diversity. *ISME Journal* 1, 283–90.
22. Taylor, D. L., et al. (2014). A first comprehensive census of fungi in soil reveals both hyperdiversity and fine-scale niche partitioning. *Ecological Monographs* 84, 3–20.
23. Soo, R. M., et al. (2009). Microbial biodiversity of thermophilic communities in hot mineral soils of Tramway Ridge, Mount Erebus, Antarctica. *Environmental Microbiology* 11, 715–28.
24. Boag, B., and Yeates, G. W. (1998). Soil nematode biodiversity in terrestrial ecosystems. *Biodiversity & Conservation* 7, 617–30.
25. Wu, T. H., et al. (2011). Molecular study of the worldwide distribution and diversity of soil animals. *Proceedings of the National Academy of Sciences* 108, 17720–5.
26. Wu, et al. (2011). Molecular study of the worldwide distribution and diversity of soil animals.
27. Fierer, N., and Jackson, R. B. (2006). The diversity and biogeography of soil bacterial communities. *Proceedings of the National Academy of Sciences* 103, 626–31.
28. Tedersoo, L., et al. (2014). Global diversity and geography of soil fungi. *Science* 346, 1078.
29. Müller, P. E. (1884). Studier over skovjord, som bidrag til skovdyrkningens theori. II. Om muld og mor i egeskove og paa heder. *Tidsskrift for Skovbrug* 7, 1–232.
30. Mawdsley, J. L., and Bardgett, R. D. (1997). Continuous defoliation of perennial ryegrass (*Lolium perenne*) and white clover (*Trifolium repens*) and associated changes in the microbial population of an upland grassland soil. *Biology and Fertility of Soils* 24, 52–8.
31. Schadt, C. W., et al. (2003). Seasonal dynamics of previously unknown fungal lineages in tundra soils. *Science* 301, 1359–61.
32. Bardgett, R. D., et al. (2007). Heterotrophic microbial communities use ancient carbon following glacial retreat. *Biology Letters* 3, 487–90.
33. Setälä, H., and McLean, M. A. (2004). Decomposition rate of organic substrates in relation to the species diversity of soil saprophytic fungi. *Oecologia* 139, 98–107.
34. van der Heijden, M. G. A., et al. (1998). Mycorrhizal fungal diversity determines plant biodiversity, ecosystem variability and productivity. *Nature* 396, 72–5.

35. Cole, L., et al. (2004). Soil animals influence microbial abundance, but not plant-microbial competition for soil organic nitrogen. *Functional Ecology* 18, 631–40.
36. de Vries, F. T., et al. (2013). Soil food web properties explain ecosystem services across European land use systems. *Proceedings of the National Academy of Sciences* 110, 14296–301.
37. Handa, I. T., et al. (2014). Consequences of biodiversity loss for litter decomposition across biomes. *Nature* 509, 218–21.
38. Blanka, V., et al. (2009). Ants accelerate succession from mountain grassland towards spruce forest. *Journal of Vegetation Science* 20, 577–87.
39. Pennis, E. (2015). Africa's soil engineers: Termites. *Science* 347, 596–7.
40. Bonachela, J. A., et al. (2015). Termite mounds can increase the robustness of dryland ecosystems to climate change. *Science* 347, 651–5.
41. Van der Putten, W. H., Vet, L. E. M., and Peters, B. A. M. (1993). Plant-specific soil-borne diseases contribute to succession in foredune vegetation. *Nature* 362, 53–6.
42. Klironomos, J. N. (2002). Feedback with soil biota contributes to plant rarity and invasiveness in communities. *Nature* 417, 67–70.
43. Mangan, S. A., et al. (2010). Negative plant-soil feedback predicts tree-species relative abundance in a tropical forest. *Nature* 466, 752–5.
44. Mangan, et al. (2010). Negative plant-soil feedback predicts tree-species relative abundance in a tropical forest.
45. Reinhart, K. O., et al. (2003). Plant-soil biota interactions and spatial distribution of black cherry in its native and invasive ranges. *Ecology Letters* 6, 1046–50.
46. Evelyn, J. (1679). *Terra, a Philosophical Essay of Earth*. London: The Royal Society.
47. Daubeny, C. (1845). Memoir on the rotation of crops, and on the quantity of inorganic matters abstracted from the soil by various plants under different circumstances. *Philosophical Transactions of the Royal Society, London* 135, 179–252.
48. Rothamsted Experimental Station, now Rothamsted Research, is the oldest agricultural research station in the world, founded by John Lawes and Joseph Gilbert in 1843.
49. Whitney, M., and Cameron, F. K. (1903). The chemistry of soil as related to crop production. *Bulletin* 22, United States Bureau of Soils.
50. Whitney, M. (190)9. *Soils of the United States*. US Department of Agriculture, Bureau of Soils Bulletin 55. Washington, DC.
51. Loughridge, R. H. (1961). The scientific work of Eugene Woldemar Hilgard. *Science* 43, 450–3.
52. Hawkesford, M. J., et al. (2013). Prospects of doubling global wheat yields. *Food and Energy Security* 2, 34–48.

53. United States Department of Agriculture (2010). National Resources Inventory, Summary Report. Natural Resources Conservation Service.
54. Montgomery, D. R. (2007). Soil erosion and agricultural sustainability. *Proceedings of the National Academy of Sciences* 104, 13268–72.
55. Jones, D. L., et al. (2013). Nutrient stripping: The global disparity between food security and soil nutrient stocks. *Journal of Applied Ecology* 50, 851–62.
56. Ju, X. T., et al. (2009). Reducing environmental risk by improving N management in intensive Chinese agricultural systems. *Proceedings of the National Academy of Sciences* 106, 3041–6.
57. Guo, J. H., et al. (2010). Significant acidification in major Chinese croplands. *Science* 327, 1008–10.
58. Ju, et al. (2009). Reducing environmental risk by improving N management in intensive Chinese agricultural systems.
59. Oldeman, L. R., et al. (1990). *Global Assessment of Soil Degradation*. Wageningen: International Soil Reference Information Centre.
60. Massa, C., et al. (2012). A 2500 year record of natural and anthropogenic soil erosion in South Greenland. *Quaternary Science Reviews* 32, 119–30.
61. Greipsson, S. (2012). Catastrophic soil erosion in Iceland: Impact of long-term climate change, compounded natural disturbances and human driven land-use changes. *Catena* 98, 41–54.
62. Baveye, P. C., et al. (2011). From dust bowl to dust bowl: Soils are still very much a frontier of science. *Soil Science Society of America Journal* 75, 2037–48.
63. Allen, V. G. (2007). Integrated irrigated crop-livestock systems in dry climates. *Agronomy Journal* 99, 346–60.
64. Weiss, H., et al. (1993). The genesis and collapse of Third Millennium North Mesopotamian Civilization. *Science* 261, 995–1004.
65. Vanlauwe, B., Six, J., Sanginga, N. and Adesina, A. A. (2015) Soil fertility decline at the base of rural poverty in sub-Saharan Africa. *Nature Plants*, 1, 15101. http://www.nature.com/articles/nplants2015101
66. Hillel, D. (1991), *Out of the Earth: Civilization and the Life of the Soil*, Berkeley and Los Angeles: University of California Press; Montgomery, D. R. (2007), *Dirt: The Erosion of Civilizations*, Berkeley and Los Angeles: University of California Press.
67. Roosevelt, Franklin D., 'Letter to all State Governors on a Uniform Soil Conservation Law', 26 February 1937.
68. The US Soil Conservation Service was renamed the Natural Resource Conservation Service in 1994 to better reflect its broadening scope.
69. Howard, Albert (1940). *An Agricultural Testament*. New York and London: Oxford University Press.
70. Faulkner, E. H. (1943). Plowman's folly. *Soil Science* 56, 394.
71. Huggins, D. R., and Reganold, J. P (2008). No-till: The quiet revolution. *Scientific American* 299, 70–7.
72. Huggins and Reganold (2008). No-till: The quiet revolution.

73. Pittelkow, C. M., et al. (2015). Productivity limits and potentials of the principles of conservation agriculture. *Nature* 517, 365–8.
74. Brooker, R. W., et al. (2015). Improving intercropping: A synthesis of research in agronomy, plant physiology and ecology. *New Phytologist* 206, 107–17.
75. Glaser, B. (2007). Prehistorically modified soils of central Amazonia: A model for sustainable agriculture in the twenty-first century. *Philosophical Transactions of the Royal Society, B.* 362, 187–96.
76. Glaser (2007). Prehistorically modified soils of central Amazonia.
77. Glaser and Birk (2012). State of the scientific knowledge on properties and genesis of Anthropogenic Dark Earths in Central Amazonia (terra preta de Índio).
78. Glaser and Birk (2012). State of the scientific knowledge on properties and genesis of Anthropogenic Dark Earths in Central Amazonia (terra preta de Índio).
79. Glaser (2007). Prehistorically modified soils of central Amazonia.
80. Glaser (2007). Prehistorically modified soils of central Amazonia.
81. The official definition of 'terroir' adopted by the International Organisation of Vine and Wine (RESOLUTION OIV/VITI 333/2010I) is 'a concept which refers to an area in which collective knowledge of the interactions between the identifiable physical and biological environment and applied vitivinicultural practices develops, providing distinctive characteristics for the products originating from this area'.
82. White, M. A. (2009). Land and wine. *Nature Geosciences* 2, 82–4.
83. Huggett, J. M. (2006). Geology and wine: A review. *Proceedings of the Geologists' Association* 117, 239–47.
84. van Leeuwen, C., et al. (2004). Influence of climate, soil, and cultivar on terroir. *American Journal of Enology and Viticulture* 55, 207–17.
85. Garcia, J.-P. (2011). Les Sols viticoles de Bourgogne: Élaboration naturelle et construction humaine. *Revue des œnologues et des techniques vitivinicoles et œnologicques: Magazine trimestriel d'information professionnelle* 38, 62–4.
86. Gougeon, R. D., et al. (2009). The chemodiversity of wines can reveal a metabologeography expression of cooperage oak wood. *Proceedings of the National Academy of Sciences* 106, 9174–9.
87. Roullier-Gall, C., Lucio, M., Noret, L., Schmitt-Kopplin, P., and Gougeon, R. D. (2014). How subtle is the 'terroir' effect? Chemistry-related signatures of two 'climats de Bourgogne'. *PloS One* 9, e97615.
88. UN-Habitat (2001). *Cities in a Globalizing World: Global Report on Human Settlements*. United Nations Centre for Human Settlements.
89. Davies, L., et al. (2011). *Urban: UK National Ecosystem Assessment*, chapter 10.
90. European Commission (2012). *Guidelines on Best Practice to Limit, Mitigate or Compensate Soil Sealing*. Luxembourg: Publications Office of the European Union.

91. European Commission (2012). *Guidelines on Best Practice to Limit, Mitigate or Compensate Soil Sealing*.

92. European Commission (2012). *Guidelines on Best Practice to Limit, Mitigate or Compensate Soil Sealing*.

93. Perry, T., and Nawaz, R. (2008). An investigation into the extent and impacts of hard surfacing of domestic gardens in an area of Leeds, United Kingdom. *Landscape and Urban Planning* 86, 1–13.

94. Haase, D., and Nuissl, H. (2007). Does urban sprawl drive changes in the water balance and policy? The case of Leipzig (Germany) 1870–2003. *Landscape and Urban Planning* 80, 1–13.

95. Kolokotroni, M., and Giridharan, R. (2008). Urban heat island intensity in London: An investigation of the impact of physical characteristics on changes in outdoor air temperature during summer. *Solar Energy* 82, 986–98.

96. Davies, et al. (2011). *Urban*, chapter 10.

97. New York City Soil Survey. http://www.soilandwater.nyc/soil.html

98. Edmondson, J. L., et al. (2012). Organic carbon hidden in urban ecosystems. *Scientific Reports* 2, 963.

99. Edmondson, J. L., et al. (2014). Urban cultivation in allotments maintains soil qualities adversely affected by conventional agriculture. *Journal of Applied Ecology* 51, 880–9.

100. Edmondson, et al. (2014). Urban cultivation in allotments maintains soil qualities adversely affected by conventional agriculture.

101. De Castella, T. (2013). How much of the UK is covered in golf course? http://www.bbc.co.uk/news/magazine-24378868

102. Golf Course Superintendents Association of America Golf Course Environmental Profile (2007).

103. De Castella (2013). How much of the UK is covered in golf course?

104. Kim, B. F., et al. (2014). Urban community gardeners' knowledge and perceptions of soil contaminant risks. *PloS One* 9, e87913.

105. Wei, B., and Yan, L. (2010). A review of heavy metal contaminations in urban soils, urban road dusts and agricultural soils from China. *Microchemical Journal* 94, 99–107.

106. Mielke, H. W., et al. (1983). Lead concentrations in inner-city soils as a factor in the child lead problem. *American Journal of Public Health* 73, 1366–9.

107. Manta, D. S., et al. (2002). Heavy metals in urban soils: A case study from the city of Palermo (Sicily), Italy. *Science of the Total Environment* 300, 229–43.

108. UK Soil and Herbage Pollutant Survey (2007). Environmental concentrations of polychlorinated dibenzo-p-dioxins and polychlorinated dibenzofurans in UK soil and herbage, Report no. 10. Environment Agency.

109. Kessler, R. (2013). Urban gardening: Managing the risks of contaminated soil. *Environmental Health Perspectives* 121, 327–33.

110. EPA (2011). Brownfields and Urban Agriculture: Interim Guidelines for Safe Gardening Practices. Chicago, IL: Region 5 Superfund Division, U.S. Environmental Protection Agency.

111. Oliver, L., et al. (2005). The scale and nature of European brownfields. In *Proceedings of CABERNET 2005: The International Conference on Managing Urban Land.*

112. The United States Conference of Mayors (2010). Recycling America's Land: A National Report on Brownfields Redevelopment (1993–2010).

113. The UK Olympic Delivery Authority (ODA), the body responsible for coordinating the delivery of the London 2012 Olympic and Paralympic Games, commissioned Atkins to carry out the remediation of the site in preparation for its development.

114. King, F. H. (1911). *Permanent Agriculture in China, Korea and Japan.* Madison.

115. http://www.manchestereveningnews.co.uk/news/nostalgia/soil-soaked-in-the-blood-of-middleton-men-1009260

116. Hupey, J. (2006). The long-term effects of explosive munitions on the WWI battlefield surface of Verdun, France. *Scottish Geographical Journal* 122, 167–84.

117. Macdonald, L. (1978). *They Called it Passchendaele.* London: Michael Joseph.

118. Macdonald (1978). *They Called it Passchendaele.*

119. Cited in Macdonald (1978). *They Called it Passchendaele.*

120. Das, S. (2005). *Touch and Intimacy in First World War Literature.* Cambridge: Cambridge University Press.

121. Theodore A. Dodge: Journal, 21–4 January 1863. In Brooks D. Simpson (ed.) (2013). *The Civil War: The Third Year Told by Those Who Lived It.* New York: Library of America.

122. http://www.ma150.org/day-by-day/1863-01-20/mud-march-begins

123. Barton, P., Doyle, P., and Vandewalle, J. (2005). *Beneath Flanders Fields: The Tunnellers' War 1914–1918.* Montreal: McGill-Queen's University Press.

124. Barton, Doyle, and Vandewalle (2005). *Beneath Flanders Fields.*

125. Barton, Doyle, and Vandewalle (2005). *Beneath Flanders Fields.*

126. Barton, Doyle, and Vandewalle (2005). *Beneath Flanders Fields.*

127. Barton, Doyle, and Vandewalle (2005). *Beneath Flanders Fields.*

128. Mangold, T., and Panycate, J. (2005). *The Tunnels of Cu Chi: A Remarkable Story of War in Vietnam.* London: Cassell.

129. Mangold and Panycate (2005). *The Tunnels of Cu Chi.*

130. The soft laterite discovered by Buchanan-Hamilton has since been found to be very rare, and potentially misleading, in that a later description of the site revealed that pickaxes were needed to excavate the material.

131. Mangold and Panycate (2005). *The Tunnels of Cu Chi.*

132. Doyle, P., et al. (2010). Yellow Sands and Penguins: The Soil of 'the Great Escape'. In Landa., E. R. and Feller, C., *Soil and Culture.* Dordrecht: Springer.

133. Hupey, J. P., and Schaetzl, R. J. (2006). Introducing 'bombturbation,' a singular type of soil disturbance and mixing. *Soil Science* 171, 823–36.

134. Hupey and Schaetzl (2006). Introducing 'bombturbation,' a singular type of soil disturbance and mixing.

135. Hupey, J. P., and Schaetzl, R. J. (2008). Soil development on the WWI battlefield of Verdun, France. *Geoderma* 145, 37–49.

136. Stellman, et al. (2003). The extent and patterns of usage of Agent Orange and other herbicides in Vietnam. *Nature* 422, 681–7.

137. Schecter, A., et al. (1995). Agent Orange and the Vietnamese: The persistence of elevated dioxin levels in human tissues. *American Journal of Public Health* 84, 516–22.

138. Kahn, P. C., et al. (1988). Dioxins and dibenzofurans in blood and adipose tissue of Agent Orange-exposed Vietnam veterans and matched controls. *Journal of the American Medical Association* 259, 1661–7.

139. Dwernychuk, L., et al. (2002). Dioxin reservoirs in southern Viet Nam—A legacy of Agent Orange. *Chemosphere* 47, 117–37.

140. Schecter, A., et al. (2001). Recent dioxin contamination from Agent Orange in residents of a southern Vietnam city. *Journal of Occupational and Environmental Medicine* 43, 435–43.

141. Depleted uranium is a by-product of the nuclear power industry and is dominantly comprised of uranium depleted in the fissile ^{235}U isotope (i.e. it is ~99.8 atom% ^{238}U). The material used in penetrators is alloyed with a small amount of titanium (0.75% Ti and 99.25% U) to improve machinability. It is known as 'CHARM3' alloy.

142. Bleise, A., et al. (2003). Properties, use and health effects of depleted uranium (DU): A general overview. *Journal of Environmental Radioactivity* 64, 93–112.

143. Bem, H., and Bou-Rabee, F. (2004). Environmental and health consequences of depleted uranium use in the 1991 Gulf War. *Environment International* 30, 123–34; Danesi, P. R., et al. (2003). Depleted uranium particles in selected Kosovo samples. *Journal of Environmental Radioactivity* 64, 143–54.

144. Alvarez, R., et al. (2011). Geochemical and microbial controls of the decomposition of depleted uranium in the environment: Experimental studies using soil microorganisms. *Geomicrobiology Journal* 28, 457–70.

145. Handley-Sidhu, S., et al. (2009). Biogeochemical controls on the corrosion of depleted uranium alloy in subsurface soils. *Environmental Science and Technology* 43, 6177–82.

146. The Comprehensive Nuclear-Test-Ban Treaty (CTBT) is a multilateral treaty that was adopted by the United Nations General Assembly in 1996, by which nations agreed to ban all nuclear explosions in all environments. Although the Treaty has yet to come into force, due to the non-ratification of some nations, only a handful of nuclear bomb tests have been done since

the CTBT was opened for signature. These include tests in India, Pakistan, and Korea.

147. Kudo, A., et al. (1991). Geographical distribution of fractioned local fallout from the Nagasaki A-bomb. *Journal of Environmental Radioactivity* 14, 305–16.

148. The half-life—or amount of time needed for its amount to fall by half—of ^{239}Pu is 24,200 years, whereas that of ^{137}Cs is about thirty years.

149. As detailed in Chapter 2, mycorrhizal fungi form symbiotic associations with most plants, whereby the plant provides the fungus with carbon, derived from photosynthesis, and the fungus captures nutrients from soil and passes them on to the plant.

150. Skuterud, L., et al. (1997). Contribution of fungi to radiocaesium intake by rural populations in Russia. *Science of the Total Environment* 193, 237–42.

151. Pollan, M. (2008). Farmer in Chief. *New York Times Magazine*, 9 October.

152. Life on the Home Front: Oregon Responds to World War II. http://arcweb. sos.state.or.us/pages/exhibits/ww2/services/ag.htm

153. *Dayton Review*, 2 August 1945.

154. Clemmensen, K. E., et al. (2013). Roots and associated fungi drive long-term carbon sequestration in boreal forest. *Science* 339, 1615–18.

155. UK Countryside Survey (2007). http://www.countrysidesurvey.org.uk/ reports-2007 NERC Centre for Ecology and Hydrology.

156. Ward, S. E., et al. (2013). Managing grassland diversity for multiple ecosystem services: Determination of the distribution and temperature sensitivity of carbon pools in English grasslands. Defra Report BD5003.

157. Edmondson, J. L., et al. (2012). Organic carbon hidden in urban ecosystems.

158. Dorrepaal, E., et al. (2009). Carbon respiration from subsurface peat accelerated by climate warming in the subarctic. *Nature* 460, 616–19.

159. Schuur, E. A. G., Vogel, J. G., and Crummer, K. G. (2009). The effect of permafrost thaw on old carbon release and net carbon exchange from tundra. *Nature* 459, 556–9.

160. Schuur, Vogel, and Crummer (2009). The effect of permafrost thaw on old carbon release and net carbon exchange from tundra.

161. Gange, A. C., et al. (2007). Rapid and recent changes in fungal fruiting patterns. *Science* 316, 71.

162. Kauserud, H., et al. (2008). Mushroom fruiting and climate change. *Proceedings of the National Academy of Sciences U.S.A.* 105, 3811–14.

163. Zhou, J., et al. (2012). Microbial mediation of carbon-cycle feedbacks to climate warming. *Nature Climate Change* 2, 106–10.

164. Melillo, J. M., et al. (2011). Soil warming, carbon–nitrogen interactions, and forest carbon budgets. *Proceedings of the National Academy of Sciences U.S. A* 108, 9508–12.

165. Lewis, S. L., et al. (2011). The 2010 Amazon drought. *Science* 331, 6017.

NOTES AND REFERENCES

166. Beniston, M., et al. (2003). Estimates of snow accumulation and volume in the Swiss Alps under changing climatic conditions. *Theoretical and Applied* 76, 125–40.

167. Peacock, S. (2012), Projected Twenty-First-Century Changes in Temperature, Precipitation, and Snow Cover over North America in CCSM4, *Journal of Climate* 25, 4405–29; Burakowski, E., and Magnusson, M. (2012). Climate impacts on the winter tourism economy in the United States. Natural Resources Defense Council.

168. Nearing, M. A., et al. (2005). Modeling response of soil erosion and runoff to changes in precipitation and cover. *Catena* 61, 131–54.

169. Mullan, D. (2013). Soil erosion under the impacts of future climate change: Assessing the statistical significance of future changes and the potential on-site and off-site problems. *Catena* 109, 234–46.

170. The Duke Forest experiment consists of four free-air CO_2 enrichment (FACE) plots that provide elevated atmospheric CO_2 concentration and four plots that provide ambient CO_2 control. This allows scientists to explore how elevated carbon dioxide affects both plants and soils under real field conditions.

171. Drake, J. E., et al. (2011). Increases in the flux of carbon belowground stimulate nitrogen uptake and sustain the long-term enhancement of forest productivity under elevated CO_2. *Ecology Letters* 14, 349–57.

172. Wookey, P., et al. (2009). Ecosystem feedbacks and cascade processes: Understanding their role in the responses of Arctic and alpine ecosystems to environmental change. *Global Change Biology* 15, 1153–72.

173. Post, E., and Pedersen, C. (2008). Opposing plant community responses to warming with and without herbivores. *Proceedings of the National Academy of Sciences U.S.A* 105, 12353–8.

174. Wolf, B., et al. (2010). Grazing-induced reduction of natural nitrous oxide release from continental steppe. *Nature* 464, 881–4.

175. Melody C., and Schmidt, O. (2012). Northward range extension of an endemic soil decomposer with a distinct trophic position. *Biology Letters* 8, 956–9.

176. Kurz, W. A., et al. (2008). Mountain pine beetle and forest carbon feedback to climate change. *Nature* 452, 987–90.

177. Powlson, D. S., Whitmore, A. P., and Goulding, K. W. T. (2011). Soil carbon sequestration to mitigate climate change: A critical re-examination to identify the true and the false. *European Journal of Soil Science* 62, 42–55.

178. Lal, R. (2004). Soil carbon sequestration impacts on global climate change and food security. *Science* 304, 1623–7.

179. Soussana, J. F., et al. (2004). Carbon cycling and sequestration opportunities in temperate grasslands. *Soil Use and Management* 20, 219–30.

180. Six, J., et al. (2004). The potential to mitigate global warming with no-tillage management is only realized when practised in the long term. *Global Change Biology* 10, 155–60.
181. Powlson, D. S., et al. (2014). Limited potential for no-till agriculture for climate change mitigation. *Nature Climate Change* 4, 678–83.
182. Rochette, P. (2008). No-till only increases N$_2$O emissions in poorly aerated soils. *Soil and Tillage Research* 101, 97–100.
183. Tilman, D., Hill J., and Lehman, C. (2006). Carbon-negative biofuels from low-input high-diversity grassland biomass. *Science* 314, 1598–600.
184. Fornara, D. A., and Tilman, D. (2008). Plant functional composition influences rates of soil carbon and nitrogen accumulation. *Journal of Ecology* 96, 314–22.
185. De Deyn, G. B., et al. (2011). Additional benefits for carbon sequestration of grassland biodiversity restoration. *Journal of Applied Ecology* 48, 600–8.
186. Kell, D. (2012). Large-scale sequestration of atmospheric carbon via plant roots in natural and agricultural ecosystems: Why and how. *Philosophical Transactions of the Royal Society B* 367, 1589–97.
187. Funded by the World Bank, the International Development Association, and the government, projects were set up to restore China's heavily degraded Loess Plateau and alleviate poverty in this region. Centuries of unsustainable farming led to this part of China suffering catastrophic soil erosion and widespread poverty. But the introduction of soil conservation and sustainable farming practices in parts of this region has revitalized the soil, which in turn has reduced sediment run-off, and increased food production and farm incomes. For further details, see *Rehabilitating a Degraded Watershed: A Case Study from China's Loess Plateau* (2010). The World Bank Institute.
188. Shangguan, W., et al. (2014). Soil diversity as affected by land use in China: Consequences for soil protection. *Scientific World*, Article ID 913852.
189. Amundson, R., Guo, Y., and Gong, P. (2003). Soil diversity and land use in the United States. *Ecosystems* 6, 470–82.
190. Tennesen, M. (2014). Rare earth. *Science* 346, 692–5.
191. Godfray, H. C. J., et al. (2010), Food security: The challenge of feeding 9 billion people, *Science* 327, 812–18; Foley, J. A. (2011), Can we feed the world & sustain the planet? *Scientific American* 305, 60–5.
192. Natural Environment Research Council (2012), Most Wanted II: Postgraduate and professional skills needs in the Environment Sector. NERC, UK.
193. Global Soil Partnership, Intergovernmental Technical Panel on Soils (www.fao.org/globalsoilpartnership/intergovernmental-technical-panel-on-soils/en/).
194. http://www.fao.org/globalsoilpartnership/highlights/detail/en/c/215220/

INDEX

COLLIDING CONTINENTS

A geological exploration of the Himalaya, Karakoram, and Tibet

Mike Searle

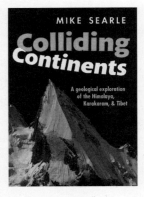

978-0-19-965300-3 | Hardback | £25.00

'There's something here to please anyone on the geology spectrum: the student wanting to understand how the fundamentals are applied; the academic intrigued by the science; the climber dreaming of virgin territory. All can learn from the master in this excellent book.' Simon Cook, *Oman Daily Observer*

The crash of the Indian plate into Asia is the biggest known collision in geological history, and it continues today. The result is the Himalaya and Karakoram—one of the largest mountain ranges on Earth. In this beautifully illustrated book, Mike Searle, a geologist at the University of Oxford and one of the most experienced field geologists of our time, uses his personal accounts of extreme mountaineering and research in the region to piece together the geological processes that formed such impressive peaks.

Sign up to our quarterly e-newsletter **http://academic-preferences.oup.com/**

THE EMERALD PLANET

How plants changed Earth's history

David Beerling

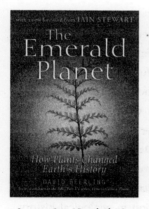

978-0-19-954814-9 | Paperback | £11.99

'My favorite nonfiction book this year. A minutely-argued but highly readable history of the last half-billion years on earth. The story Beerling tells could not have been put together even ten years ago, for it depends upon the latest insights from palaeontology, climate science, genetics, molecular biology, and chemistry, all brilliantly and beautifully integrated together. I got a special deep, quiet pleasure from reading *The Emerald Planet*—the sort of pleasure one gets from reading Darwin.' Oliver Sacks,

Book of the Year, *Observer*

Plants have profoundly moulded the Earth's climate and the evolutionary trajectory of life. David Beerling puts plants centre stage, revealing the crucial role they have played in driving global changes in the environment, in recording hidden facets of Earth's history, and in helping us to predict its future.

Sign up to our quarterly e-newsletter **http://academic-preferences.oup.com/**

THE GOLDILOCKS PLANET

The 4 billion year story of Earth's climate

Jan Zalasiewicz and Mark Williams

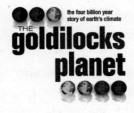

JAN ZALASIEWICZ & MARK WILLIAMS

978-0-19-968350-5 | Paperback | £10.99

'A balanced, well written, mostly comprehensive and well-argued book.'

Times Higher Education Supplement

In this remarkable new work, Jan Zalasiewicz and Mark Williams demonstrate how the Earth's climate has continuously altered over its 4.5 billion-year history. The story can be read from clues preserved in the Earth's strata—the evidence is abundant, though always incomplete, and also often baffling, puzzling, infuriating, tantalizing, and seemingly contradictory. Geologists, though, are becoming ever more ingenious at interrogating this evidence, and the story of the Earth's climate is now being reconstructed in ever-greater detail—maybe even providing us with clues to the future of contemporary climate change.

OCEAN WORLDS

The story of seas on Earth and other planets

Jan Zalasiewicz and Mark Williams

978-0-19-967288-2 | Hardback | £20.00

'Fluid and fascinating prose with just the right dosage of entertaining anecdotes and human interest' Michael Gross, *Chemistry & Industry*

'Readable and absorbing account' *The Guardian*

Oceans make up most of the surface of our blue planet. As climate change, pollution, and over-exploitation by humans put this precious resource at risk, it is more important than ever that we understand and appreciate the nature and history of oceans. There is much we still do not know about the story of the Earth's oceans, and we are only just beginning to find indications of oceans on other planets.

Jan Zalasiewicz and Mark Williams consider the deep history of oceans, how and when they may have formed on the young Earth—topics of intense current research—how they became salty, and how they evolved through Earth history. They look at clues to possible seas that may once have covered parts of Mars and Venus and that may still exist, below the surface, on moons such as Europa and Callisto, and the possibility of watery planets in other star systems.

THE PLANET IN A PEBBLE

A journey into Earth's deep history

Jan Zalasiewicz

978-0-19-964569-5 | Paperback | £9.99

'A mind-expanding, awe inducing but friendly scientific exploration of the history' Holly Kyte, *The Sunday Telegraph*

This is a narrative of the Earth's long and dramatic history, as gleaned from a single pebble. It begins as the pebble-particles form amid unimaginable violence in distal realms of the Universe, in the Big Bang, and in supernova explosions, and continues amid the construction of the Solar System. Jan Zalasiewicz shows the almost incredible complexity present in such a small and apparently mundane object. It may be small, and ordinary, this pebble—but it is also an eloquent part of our Earth's extraordinary, never-ending story.

WAKING THE GIANT

How a changing climate triggers earthquakes, tsunamis, and volcanoes

Bill McGuire

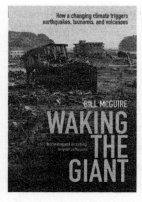

978-0-19-967875-4 | Paperback | £11.99

'McGuire traces this fascinating and disturbing story from the past in order to alert us to present and future perils'

Geographical Magazine

The last twenty thousand years has seen our planet flip from icehouse to greenhouse, provoking earthquakes, tsunamis, and volcanic outbursts. Like a giant stirring from a long sleep, the Earth beneath our feet tossed and turned. Some fifteen thousand years ago, ice sheets kilometres thick buried much of Europe and North America, and sea levels were lower. The following fifteen millennia, however, saw an astonishing transformation as our planet changed into the temperate world upon which our civilization has grown and thrived. In *Waking the Giant*, Bill McGuire argues that climate change is once more setting the scene for the giant to reawaken. Are we leaving our children not only a far hotter world, but also a more geologically fractious one?

Sign up to our quarterly e-newsletter http://academic-preferences.oup.com/